PLANET WITHOUT APES

PLANET WITHOUT APES

CRAIG B. STANFORD

The Belknap Press of Harvard University Press

CAMBRIDGE, MASSACHUSETTS

LONDON, ENGLAND

2012

*Dedicated to the many individuals around the globe
who have devoted their lives to the protection and
conservation of the last remaining great apes*

Library of Congress Cataloging-in-Publication Data
Stanford, Craig B. (Craig Britton), 1956–
Planet without apes / Craig B. Stanford.
p.   cm.
Includes bibliographical references and index.
ISBN 978-0-674-06704-2 (alk. paper)
1. Apes   2. Endangered species.   3. Extinct animals.   I. Title.
QL737.P96S733   2012
599.88—dc23   2012023985

Contents

Save the Apes!

There are seven billion people on Earth today. Meanwhile, the population of our next-of-kin is plummeting to extinction. After millions of years of co-existence with humans, they have been nearly exterminated within a few decades and seem destined to go the way of the American bison, the giant panda, and the tiger; reduced to such pathetically low numbers that they exist only in carefully managed, protected areas. As in other genocides, the world watches, wrings its hands, but does very little to stop it. The result of the slaughter will be that the great apes, our closest relatives on Earth, will be effectively gone. They will hardly be alone; there is a mass extinction afoot such as our planet has never seen. There have been at least five mass extinctions in Earth's history. During each a significant portion of the plants and animals disappeared. The biologist Stuart Pimm and his colleagues estimate the current extinction rate to be between one hundred and one thousand times greater than the rate of species disappearance that has characterized the past half-billion years. In other words, we are in the midst of the sixth mass extinction, and it has barely begun.

Given that all animals are at risk from our rape of the Earth, why should we be so concerned about just four species—the chimpanzee, gorilla, bonobo, and orangutan—among the many millions alive today? We should care because these four creatures are our lifeline. They are our last remaining links to our evolutionary past, and they tell us a great deal about who we are today. Allowing them to die would be like allowing our extended family to die. But we are not allowing them to perish through benign neglect. Humans have carried out a campaign of extermination against the great apes that has reached epic proportions. It has pushed two of the ape species to the brink of extinction in the wild, with the remaining two not far behind.

Great apes have the tragic misfortune to live only in the most impoverished regions of the Earth. Their tropical forest homes are surrounded by people living in abject poverty, for whom protecting a rare animal is inconsequential compared to feeding a family. The apes live long lives, maturing slowly and reproducing at a very slow pace, and so are unable to withstand the panoply of threats that now envelop them. Their forests are being cut down beneath them. Their meat is relished by people. They are subject to all the diseases that afflict humans and have suffered massive epidemics from emerging viruses. They continue to be taken from the wild to serve as laboratory animals, circus performers, and household pets. After millennia of co-existing with people, the great apes face an onslaught of trouble that has caused them to plummet toward extinction across their ranges in Africa and Asia.

Why should we care? That is the question that conservationists have tried to answer for the public for generations, whether the cause was Save the Whales, Save the Rain Forests, or save anything else. Perhaps the question we need to ask in this case is why we should care more about the great apes than we might care about the myriad other animals in need of our protection. After all, a gorilla is not as beautiful as a tiger, or as charismatic as a giant panda, or as important to global ecosystems as honey bees. We care about the great apes first and foremost because they are us. They are largely the same genetic material of which we are made—our evolutionary closest kin. They are like humans in ways that blur the distinctions between us.

How intelligent is an ape? There is a chimpanzee in a lab in Japan who can remember numerical sequences better than you or I can. There are chimpanzees who talk to themselves, making signs with their fingers as they absent-mindedly leaf through a magazine or chat with each other. There is a bonobo who has learned to make a stone tool in much the same way our ancestors did, and who has come to treat the best tools he has made as valued creations. A chimpanzee in a zoo in Sweden collects objects in his enclosure in the hours before the zoo opens in order to have a stockpile of projectiles to hurl at annoying zoo visitors whom he knows will arrive late in the day.

Such displays of intellect are most obvious in captive apes reared by humans, but in the wild these animals show off powers of humanlike reasoning too. I once watched a chimpanzee in Gombe National Park, Tanzania, as he searched for fruit in a large tree. As he climbed he passed a hole in the tree and, reaching in with his hand, flushed a woodpecker that flew out past him. He thrust his fingers into the hole in search of something, but the hole was too small to admit more than just a couple of fingers at a time. He promptly descended to the ground, selected a sizeable branch, and returned to the hole. Jabbing the branch deep into the cavity, he then gingerly extracted it dripping with egg yolk, which he licked off with relish. Anecdotes of this sort are common among scientists who study wild chimpanzees. They are the most technologically accomplished nonhuman animals on Earth by a wide margin.

Chimpanzees and the other great apes learn their technological skills by watching their peers and their families. These traditions are passed down from one generation to the next. In other words, they learn and store knowledge in much the same way that early humans did. Chimpanzee cultures have been the focus of systematic studies for years, and the evidence is plain that chimpanzee traditions are simpler versions of human cultures. When a chimpanzee population is decimated by the felling of its forest or is exterminated by hunting, we lose a culture, not just a gene pool. Only among these apes and very few other animals on Earth do these concerns apply.

The threats faced by the great apes are the focus of this book. Each chapter details a threat that has converged with other threats to push these relatives of ours to the brink. Chapter 1 lays out the problem and describes the four species of great apes. The chimpanzee, bonobo, gorilla, and orangutan are nations unto themselves, to paraphrase the writer Henry Beston, and we had better understand them on their own terms if we are going to save them. Chapter 2 describes the elephant in the room for all wildlife on Earth today: the loss of habitat. The planet loses a large chunk of its natural lands every day, and with each acre of terrain converted into farmland and towns comes the extinction of local animals and plants. Great apes are on the receiving end of habitat loss, and its rate is increasing, not decreasing.

Chapter 3 describes the bushmeat crisis, the industry of consuming great apes for food in parts of Africa. Forests that look pristine save for a few logging roads are in fact being emptied of their large animals by hunters, who sell the meat as a delicacy to people in towns and cities. Chapter 4 examines a threat little known to most of us. Ebola, anthrax, and other diseases have taken a terrible toll on wild apes across central and western Africa. Sometimes these diseases have spread from human populations to apes, and in other cases human diseases—

HIV/AIDS for one—appear to have spread in mutated form from ape to human populations. Chapter 5 is an account of the state of great apes in captivity. I examine the ethical and conservation issues involved in maintaining the thousands of apes who are in zoos and labs across the United States and Europe today.

Chapter 6 is about ecotourism. What could be wrong with ecotourism involving the great apes? In some cases, nothing at all—it can be a boon to conservation efforts. In other cases, however, badly administered ecotourism can result in additional health risks to the apes.

I have no patience for conservation accounts that portray the future of endangered wildlife in absolutely bleak terms. So each chapter offers at least a glimmer of hope for the ways in which even the most sinister threats to the great apes could be lessened or eliminated. Chapter 7 describes the recent finding that cultural variation is similar among chimpanzees to cultural diversity among ourselves, albeit on a much simpler scale. Finally, the Epilogue describes the efforts by a global community of researchers and conservationists to preserve and protect the remaining thousands of wild great apes on two continents. The fate of the apes is ultimately in the hands of these conservationists, particularly those working in the host countries in which apes live.

Planet Without Apes focuses our attention on the very real possibility of the extinction of the great apes. The idea for writing such a book grew out of many years spent studying them in remote places and seeing the human threats to their existence firsthand. Great apes have the deck stacked against them. If they are to survive to the twenty-second century and beyond, we will have to act swiftly and make some fundamental changes in how we view land use and the ethics of captive animals.

ONE

Heart of Darkness

Up a long, winding muddy river from the sea, a massive slaughter is taking place. It's happening largely out of sight from anyone who might be motivated to do something to stop it. If it were a slaughter of human beings it would be called by its rightful name: genocide. It spans the wide equatorial belt of the African continent, with a parallel slaughter being carried out half a world away in Indonesia. It has been going on for many decades, but its pace has quickened recently, and the slaughterers seem almost hell-bent to rid the world of their victims. It is carried out by the means that conquering peoples have used for centuries. The victims are shot, pierced by arrows. They are ravaged by biological weapons let loose in their midst. Mothers are shot dead, their screaming children pried from their lifeless arms to be sold to people who desire them. Sometimes they are taken captive instead of being killed outright. They live out their lives, often under abhorrent conditions, in the service of their captors. Their natural homes are burned down, chopped down, and bulldozed in the name of progress. Some activists talk about granting the victims the same basic human rights we all possess, to live free from the threat of wanton cruelty. But these are voices in the wilderness, powerless compared to the political and industrial forces at work around them, so the slaughter continues. The victims are outnumbered and out-gunned by their torturers, and in the end their fate may lie in being nostalgically remembered by the children and grandchildren of those who did the killing.

The "genocide" I am talking about is that of the last living great apes. Like the European colonists of the tropics who encountered widespread indigenous civilizations but declared the land to be "empty," those who carry out the ape genocide today do it blithely, without considering their actions a violation of any natural law. Like all colonists, we kill in the name of progress and denigrate the victims to rationalize the genocide. After all, they are animals, we are humans. In fact, we are but human animals and they are close enough to the human family to make a line we like to think of as bold rather fuzzy.

In the nineteenth and early twentieth centuries, the pioneering anthropologist Franz Boas was the first public figure to reject the idea of the "savage." Dark-skinned people in other, less technologically advanced cultures were simply that.

Any implication that their cultural milieu made them "primitives" was simply racist.[1] Just as humans did not evolve from chimpanzees (instead, both chimpanzees and humans share the same common ancestor, from which we are equally distant), western people did not evolve from nonwestern people. All people share a common and very recent ancestor. Modern cultures can be understood in the context of their environments, not in some kind of racist hierarchical ladder that puts white people on top and people of color on lower rungs. "Savages" were a romantic invention of the western world. When explorers and colonists encountered people with strange customs and languages, it was far easier to exploit them or to ignore their fundamental human rights if one first dehumanized them. As the madman Kurtz says in Joseph Conrad's *Heart of Darkness*, "Exterminate the brutes!"[2] If they weren't killed outright, they were seen as novelties to be placed on display like museum specimens or zoo animals, or "civilized" by missionaries and others, who would convert them to fit a western mold.

Savages no longer exist. Modern morality grants all people the same basic human rights, although that doesn't mean that native peoples are necessarily treated equally. Today indigenous native peoples from the Arctic to the Amazon are accorded human rights and given legal protections by governments, even as their way of life rapidly disappears. Taking their place as the dispossessed and dismissed are the great apes.

FOUR SIBLINGS

Great apes are the new "others." They stand in for the brute side of ourselves, and we see in them many of the traits we falsely attributed to indigenous people for centuries. They are the last true savages, and we are treating them about the same way we treated our human brethren for so long. We destroy their forest homes and, in doing so, exterminate not just a gene pool but also an indigenous culture, for as we have learned in recent years, cultural traditions are defining qualities of the great apes. We still use them in biomedical research, injecting them with our diseases, then locking them in tiny cages to see how the diseases progress. Some of us continue to eat them, preferring their meat to beef or chicken. In the early twenty-first century we still consider the great apes to be expendable commodities. Attitudes are changing, but it may be way too late.

Many of the ethical debates about great apes today closely echo those for indigenous people a century ago. How intelligent are they? Should they be accorded human rights? Is it cruel to put them on display in a zoo? It is difficult to imagine that we thought of other people this way only a century ago. Consider that fifty years from now we may well also consider it barbaric to imprison a gorilla in a zoo or to rear a chimpanzee in a laboratory for the sake of some psychologist's research career. Some of the developed nations—the United Kingdom,

Great ape species recognized by most authorities in 2011. All population size estimates are rough, and certainly in rapid decline. Figures based in part on IUCN Redbook (2011).

CHIMPANZEE (*Pan troglodytes* and possible subspecies). Estimated total population: 150,000–250,000

BONOBO (*Pan paniscus*). Estimated total population: 25,000–30,000 (speculative)

EASTERN GORILLA (*Gorilla beringei* and subspecies). Estimated population of mountain gorilla subspecies in Virunga National Park and Bwindi Impenetrable National Park: 750. Estimated population of eastern "Grauer's" gorilla subspecies: 5,000–8,000 (speculative)

WESTERN GORILLA (*Gorilla gorilla* and subspecies). Estimated population of western lowland subspecies: 80,000 (speculative). Estimated population of Cross River subspecies: 300

BORNEAN ORANGUTAN (*Pongo pygmaeus* and subspecies). Estimated total population: 45,000–65,000

SUMATRAN ORANGUTAN (*Pongo abelii*). Estimated total population: 7,000

Estimated Global Population of Wild Great Apes in 2011: Approximately 300,000 to 400,000

New Zealand, Germany, and others—were quick to ban the use of great apes in laboratory experimentation. These countries considered ape cognitive abilities to be so close to ours that experimentation is torture.[3] The United States has been unwilling to follow suit. To what extent do we want to extend fundamental human rights to nonhuman animals? We will take up this question in Chapter 5.

Why, in the midst of a thousand humanitarian concerns, should we care? Three million people have died in civil wars in central Africa over the past decade, and the western world doesn't seem too concerned. Homeless people die in the streets of western cities every day, and families are forced from their homes by an ailing economy. Epidemics rage, some glamorous and unsolved but others so pedestrian and simple—diarrhea remains the greatest killer of the developing world—that cleaning our water supply could save millions of lives. And yet I am asking that we give our empathy to these animal relatives. The best reason to see that this is not misanthropy is that the line between human animal and nonhuman animal—our ape brethren—is blurry.

These are the most intelligent animals in the forest, their brains outsized in relation to their bodies. They and we share a common ancestor who lived no more than a few million years ago—a blink of evolutionary time. Of the tens of millions of species that have lived and died on planet Earth, these are our four next of kin. The chimpanzee and bonobo are our true siblings; we three species split from the same tree of life about six million years ago in Africa. A couple of million years earlier, our lineage had split from that of the gorilla, also in Africa. And sometime not so long before that, the only Asian ape, the shaggy-haired orangutan, had diverged from the line. Thus, a lineage of great apes and one of humans emerged. The human lineage—the hominins—progressed through a branching tree of primitive creatures who looked like apes in some ways, such as brain size and teeth, but who resembled humans in others: they were upright bipeds who were learning to make and use tools and weapons. At times there were multiple forms of our ancestors living at the same time, sometimes even sharing the same environment. Eventually, though, only one twig on our family tree remained, which became us, *Homo sapiens*.

Meanwhile, the other fork of that great lineage diversified and was winnowed by time and natural selection. In the end four great ape species remained. Long before the divergence between hominins and apes, there had been a plethora of ape varieties, living all across Africa and Asia and into southern Europe. They slowly died off without leaving descendants, and only the more lowly monkeys and the larger brained hominins were left standing. The chimpanzee, bonobo, gorilla, and orangutan are, therefore, the modern tips of great lineages themselves. They are emphatically not, as so many people believe, primitive versions of humans, or the species from which we evolved. They are creations of evolution unto themselves and were doing quite well until the human population recently began to explode and push them inexorably off the face of the Earth.

The lives of great apes are not so different from our own. After a nearly nine-month pregnancy, a mother chimpanzee gives birth to a single baby, who is utterly dependent on her for several years. Even after being weaned from mother's milk, the baby is psychologically dependent and is unable to cope if his or her mother dies. If the baby is female, at around age eleven she will reach puberty and begin to experience sexual swellings, the pink fluid-filled appendages that inflate around the time of ovulation during each cycle. Then, at about age fifteen, she conceives and gives birth for the first time. Thereafter she gives birth about once every four years until her late forties, by which time she is elderly and in declining health. In the wild she will very likely die of disease or predator attack before age fifty, although in captivity reaching her sixties is very possible.

Sound familiar? Except for the ultimate lifespan, the life cycle of this female is very much like our own. Her reproductive potential is lower than that of a human female, and she will not undergo a marked menopause toward the end of her childbearing years. But she will live for decades, taking many years to develop from infancy to maturity. Great apes have extended kin networks that are much like the societies in which our ancestors recently lived.

NEXT OF KIN

Chimpanzees are the best known of the great apes thanks to a handful of field studies in Africa that have been going on for decades, plus a plethora of studies in captivity. Beginning with field studies in the early twentieth century that lasted only weeks (and saw little more than the fleeing rear ends of terrified chimpanzees), we have tried to penetrate their societies to understand their inner lives. Jane Goodall was, of course, the pioneer. Instead of blundering around in the forest with porters and guides and shotguns like her male predecessors, Goodall spent countless hours alone at Gombe Stream National Park, Tanzania, attempting to accustom the apes to her presence. She hoped she would eventually be allowed to get close enough to observe the details of their lives. Her strategy paid off spectacularly, and she got a closer view of what wild chimpanzees are all about than anyone at the time could have dreamed.

Goodall recorded lifelong kin bonds between mothers and their daughters. She watched males engage in Shakespearean conspiracies of subterfuge and deceit against perceived rivals. Today, chimpanzees are the best known of the great apes.[4] They live in complex polygynous societies we called fission-fusion, in which there is no long-term stable social unit except a mother and her child. All other associations are ephemeral and opportunistic. Chimpanzees come together and part ways all day long in search of their preferred food, ripe fruit, which is widely scattered in the forest. Females are sexually interested in males during their ten-day (or so) period of potential ovulation. They grow a nacreous pink, fluid-filled swelling on their rear ends that is riveting to the opposite sex. During these periods parties of many males and females may travel together, but at other times they live widely scattered lives in parties of twos or threes. The sum of all temporary parties forms the community, which may be from twenty to one hundred twenty individuals. Males form the stable core of the community, being born and raised in the same place and never leaving, whereas females emigrate at or after puberty, settling into neighboring communities to breed and raise their offspring. Males' lifelong association with each other is likely why they tend to cooperate in the course of hunting, patrolling territory, and ganging up on other males. Females don't form tight alliances with other females much at all.

There are a handful of studies of more than two decades duration across Africa, and we now understand that these chimpanzees are the only primates other than ourselves who invent novel technological solutions to everyday survival problems, and that they pass these learned traditions from generation to generation. Goodall first observed a chimpanzee selecting a twig, fashioning it into a probe, and inserting it gingerly into an earthen termite mound. When withdrawn, the stick was coated with angry termites, which the chimp proceeded to eat like cotton candy. Chimpanzees in western Africa collect rocks and clubs to use as hammers on hard-shelled nuts.[5] In the forests of central Africa, they will use large branches to pound open termite mounds, placing their feet on the branch for added power in the same way that you use a shovel in your garden.[6]

There is also a dark side to the kinship and creativity. Chimpanzees wage something resembling warfare, in which males of a community patrol their territorial boundaries, attacking and sometimes killing neighbors. They penetrate deep into enemy territory in search of neighboring males to attack. The ensuing conflicts can be horrifically violent, resulting in the brutal murder of males and sometimes females from rival communities.[7] On very rare occasions males will also gang up to attack rivals within their community.[8] Males routinely brutalize females, either through the threat of intimidation or by outright beating.

Chimpanzees are omnivores. They lust after a meaty dinner of other mammals as much as any beef-loving human. Hunting parties will spend hours chasing down, killing, and devouring monkeys, piglets, antelope fawns, and sundry small game. This discovery, made also by Goodall in 1960, was as stunning as any other. Some experts claimed meat-eating must be pathological behavior brought on by Goodall's presence in the forest. But as soon as chimpanzees were studied elsewhere, hunting was seen to be a systematic, routine part of their lives. Males do most of the hunting, and they act in concert to capture their prey, which is then divvied up among the hunters and others present according to Machiavellian rules of nepotism, manipulation, and sexual conquest. Allies get meat, rivals are

snubbed, and desired females are given meat, sometimes in exchange for sex.[9]

Most of the discoveries credited to Jane Goodall were also
observed just a few years later in the second long-term field
study of chimpanzees, in the Mahale Mountains of Tanzania.
Mahale is only sixty miles or so from Goodall's Gombe. The
famed Japanese primatologist Toshisada Nishida and a long line
of his colleagues and students have studied chimpanzees there
since the mid-1960s. As beautiful a landscape as Gombe is, with
its narrow valleys and cascading streams flowing into Lake Tanganyika, Mahale is even more stunning. It is lush and mountainous, and still has some of the large African mammals such
as buffalo and lions that are long extinct in the smaller Gombe.
Mahale chimpanzees are also adept tool users, hunters, and
at times, warriors.[10] Goodall's and Nishida's field research
have defined the boundaries of our knowledge of chimpanzees
equally. Goodall began her study a few years earlier and bene-
fited from a global English language media and a prime public
outlet in *National Geographic* magazine and television.

The endangered status of chimpanzees today belies their
incredible environmental versatility. You can find chimpanzees
in the patches of forest that dot the grasslands of parts of East
Africa. Travel westward into the heart of the vast Congo Basin
and you will find their last stronghold in the carpet of lowland
rain forests that stretch thousands of square miles. Travel farther west and you will see them in the few remaining bits of

rain forest in western Africa. And travel still farther, and having crossed nearly from the east to west coast of this vast continent, you will meet chimpanzees again in the open arid grasslands of Senegal. Recent genetic research suggests that there are in fact three distinct species of chimpanzees couched in the one species we currently know by that name.[11] In the Ruwenzori Mountains of Uganda chimpanzee nests have been spotted almost at tree line, in places so raw and cold you or I would shudder at the thought of spending a night there without a down sleeping bag. On the plains of Senegal they tolerate daytime temperatures that reach almost Death Valley levels, taking dips in pools of water to get relief from the furnace heat. For a span of five thousand kilometers the species copes with extreme equatorial environments and thrives. Except where the hand of humans intervenes.

THE SAVAGE BEAST

Gorillas were the savages the early European explorers expected to find on the shores of the African continent. The image took hold and was part and parcel of the "darkest Africa" mindset that persisted for a century or more. Brief expeditions penetrated gorilla habitat in places like the Virunga Volcanoes, but travelers brought back only anecdotes of fleeting encounters with massive shy apes. The image of the savage gorilla was only dispelled after the biologist George Schaller spent a year in the Virunga Volcanoes patiently making contact with groups, seeing them at close range for the first time. He reported timid, placid primates wanting only to be left to themselves to forage for their daily salads of fibrous veggies. Not long after Schaller's study, Dian Fossey began her pioneering work on mountain gorillas, emulating Goodall's method of settling in for the long haul, habituating the apes to her presence, then immersing herself in their daily lives. Fossey's life ended in tragedy when she lost hold of the fact that she was a guest researcher in a host nation, Rwanda, and began to take actions to protect "her" gorillas that were almost as desperate as the acts local poachers were taking against the gorillas. After a long-running series of tragic gorilla deaths, followed by Fossey's threats to retaliate against the local people, she was murdered. Soon after, her former students approached the Rwandan government to propose a new approach to gorilla research that included opening the area to ecotourism. With a new economic incentive to protect the apes, gorilla research entered a new and ultimately far more productive chapter.

Fossey's work had revealed a society very unlike that of chimpanzees: cohesive groups of a dozen or two featuring one or two massive silverback males weighing upward of two hundred kilograms.[12] These males are not as solidly in charge of the group as they appear to be at first glance. They're huge and enormously powerful and they beat their chests, but they also live in a state of perpetual worry that their females will desert them for the silverbacks of other local groups. Whereas chimpanzee societies are fluid and promiscuous, those of gorillas are cohesive. The massive apes spend most of their waking hours foraging, resting, and socializing within a few yards of one another. Indeed, researchers believe that most of a gorilla's vocal repertoire consists of close-proximity contact calls meant to keep the group informed about each member's whereabouts in dense forest undergrowth.

A certain view of gorillas and their societies was generated by the earliest field study. Dian Fossey did her pioneering work in the Virunga Volcanoes, in high mountain meadows where the air was cold and damp, and the availability of fruits and other high-energy foods limited by the climate. She reported a diet of almost exclusively foliage, much of it highly fibrous, low-calorie plants. Her gorillas were exceedingly sedentary, bulldozing their way through only a few hundred meters of forest per day. But the lives of her gorillas in the mist, it turned out, did not closely resemble gorillas elsewhere. Gorillas are found all across equatorial Africa, from primary rain forest to patches of secondary, degraded forest. Today they occur only in remnants of their former range and are persecuted by the many threats we will examine shortly. But in only a couple of clearly marginal bits of habitat in the mountains of eastern Africa do gorillas live as Fossey's did. As studies were begun elsewhere, a different picture of gorilla life emerged: this highly mobile ape preferred fruit to leaves and readily climbed tall trees to harvest the foods of choice. Gorillas are, in other words, more similar to chimpanzees in their lifestyle than Fossey ever imagined. They just don't get the chance to reveal this unless you are watching them in the habitat in which most of their numbers live.

Today we have studied gorillas across Africa from east to west. We see them as a much more versatile species than ever before, and we recognize that they often live in areas where chimpanzees cannot, because of the larger apes' ability to live on foods that are widely available but not especially nutritious. Whereas chimpanzees will not be found in a forest that lacks an abundant supply of ripe fruit, gorillas can make do by falling back onto a leafy diet for a portion of the year. This explains the presence of gorillas high in the cold Virunga Volcanoes, where neither fruit trees nor chimpanzees could survive. But as we will see, the vast heart of the gorilla's range is also a killing zone, and gorilla meat is high on the menu of many African people.

Although western and eastern gorillas look very much alike, they are genetically quite distinct, and most experts consider them separate species, further subdivided into as many as four races, or subspecies. Western lowland gorillas are the familiar zoo gorillas, their close-cropped gray hair and chestnut crowns contrasting with their mountain kins' shaggy jet-black pelage. Some authorities split lowland gorillas into multiple species rather than simply racial variants; one small and isolated relict population in a corner of Nigeria, the Cross River gorillas, appears to merit species status based on genetic studies.[13] In the east there are mountaintops that have became biological islands as a result of forest loss in the lowlands, and some host distinctive gorilla populations. Depending on how finely you like to split hairs (and perhaps how desperately you want to become known for naming a new gorilla variety), some of these may merit their own taxonomic categories.

Mountain gorillas currently number about seven hundred fifty, divided into two small forest reserves straddling the borders of Uganda, Rwanda, and the Democratic Republic of the Congo. This mountainous region, known to biologists as the Albertine Rift, features an extraordinary diversity of animal and plant life, much of it found nowhere else in Africa. It's also a region of the most sinister and chronic political instability and human suffering of any spot on Earth. If the eastern Congo Basin and the Albertine Rift were of more strategic importance to American national security than Iraq or Afghanistan are, then we would hear all about the Congo on the nightly news. Millions have died over the past decade, and they continue to die or flee from the violence that is as endemic to the region as any rare animal.

THE APE FROM VENUS

Imagine you're sitting in an African forest, watching animals go about their business. They eat, they groom one another, and mostly they rest and sleep. Then one animal approaches another and the two of them turn their torsos to face each other and, looking intermittently in each other's eyes, they have sex. What's more, the two animals are both females, and they're getting obvious intense pleasure from frantically rubbing their private parts against each other. At one point they shift grace-fully from a sitting to a prone position, one female lying on the other, still pushing her genitals back and forth, back and forth, against the other's. It would be hard for anyone not to feel a slight discomfort on first exposure to this, as human as it seems. Indeed the animals, bonobos, are the only mammals other than humans who routinely engage in sex for reasons that have noth-ing to do with reproduction.

We still know far less about bonobos than we do about chimpanzees. In fact our current state of knowledge about this ape is similar to what we knew about chimpanzees in the 1970s. This is because bonobos were discovered by western science only in the early twentieth century, in the former Belgian Congo. Unlike the ecologically versatile chimpanzee, bonobos are found only in a restricted area of the Congo Basin of central Africa, south of the bend of the great Congo River. They are also found in a far narrower range of habitats; their home is al-most always primary rain forest, with a few populations stray-ing into regions of broken forest and grassland. This fact of distribution, combined with the remoteness and chronic civil strife of their range, has made it nearly impossible to amass a decent database of information about them.

This lack of access to wild bonobos did not stop some re-searchers from exploring their behavior in captivity, and thereby hangs a tale. Although two field studies of wild bonobos were begun in the jungles of the Congo in the 1970s, most of the in-formation and photographs published about the apes in those

early years came from the San Diego Zoo. There, the primatologist Frans de Waal observed an ape so much like a chimpanzee superficially, but as observed by de Waal, utterly different in nature. Bonobos, he reported, were apes from Venus.
They engaged in sex in social contexts that often had nothing
to do with procreation and, even more impressively, in a variety
of positions that rivaled human lovemaking. Unlike the sometimes brutally male-dominated chimpanzee society, female
bonobos formed alliances that were self-empowering. Indeed, a
male bonobo who tried to brutalize a female might find himself set upon by a gang of females intent on doing him bodily
harm.[14]

De Waal also reported that, unlike chimpanzees, bonobo
sexuality was linked to a female's frequent receptivity. Instead
of a limited period of the month during which females are willing to mate, female bonobos are sexual during a more pro-
tracted period of swelling. This, combined with the fact that
bonobos travel in larger parties than chimpanzees typically do,
means that there is always a sexually receptive female bonobo or
two present. And to cap off this view of the apes as alternative
and progressive, the initial reports from the Congo indicated
that they did not share the chimpanzee's love of meat, either.
They hunted monkeys but didn't eat their prey, instead playing
with them and releasing them worse for wear but intact.

This combination of factors led de Waal to brand bonobos
as the feminist, pacifist, vegan, Kama Sutra specialists of the
nonhuman primates. The notion of the bonobo as a female-

empowered, sexually liberated pacifist stuck, and even entered the tabloid media. While this depiction seems to have merit in captivity, it doesn't square well with what we eventually learned about how bonobos behave in African forests. Although wild female bonobos do form long-lasting bonds and use them to protect themselves from male assaults, the species consumes meat avidly and doesn't mate any more frequently in the natural habitat than wild chimpanzees do.[15] In fact, recent captive studies seem to show that the hypersexuality of bonobos in zoos occurs mainly among newly constituted colonies that use sex to sort out social relationships with their new cage mates. Thereafter, frequent sex is mainly a feature of the behavior of juveniles, not adults.[16]

These academic debates about the nature of the bonobo aside, the species certainly forms a valuable counterpoint to the chimpanzee as a model for what our ancestors may have been like. More important, their status in the wild is very tenuous, living as they do in areas that have seen civil war and are subject to poaching. Even the protected populations that have been long studied are no longer safe, since when political instability forces researchers to leave the area, poachers move in and kill the apes as bushmeat.

THE RED APE

The least known of the great apes is also perhaps the most distinctive of all nonhuman primates. In the forests of the Indonesian islands of Sumatra and Borneo live enormous red-haired apes with shaggy manes. Their awkward gait during infrequent travel on the ground belies their grace and agility high in the rain forest canopy. Orangutans have been known to science for hundreds of years, but never half as well as chimpanzees and gorillas. They survived a long colonial history of the Dutch in Indonesia, who shot them and sent their carcasses to museums for study. Despite a more recent history of excellent field research, orangutans have not yielded the secrets of their lives very willingly. They are often labeled as enigmatic because of their fairly solitary nature.

Unlike the fluid communities and parties of chimpanzees and bonobos, or the tighter cohesive groups of gorillas, orangutans are very often all alone. Adult females are normally territorial and occupy patches of forest on which they tolerate no other orangutans except the occasional male, plus her offspring. They spend their lives traveling the rain forest canopy, foraging for fruits and raising young. The glacial reproductive rate of the orangutan is one of reasons the species is in such terrible trouble in the twenty-first century. Females give birth once every seven or eight years starting in their teens, which amounts to no more than a few descendants produced during a lifetime.

Males, meanwhile, spend their lives hoping to maintain a territory on which they can have regular access to females. They do this in one of at least two ways. Large adult males may occupy vast tracts of forest that encompass the ranges of several females and patrol the area to ensure that other males do not trespass. Male orangutans are bulky and slow, however, and younger, smaller males are able to sneak in while the resident is off patrolling elsewhere. We recently learned that the transient males who have been unable to control their own territory possess a fascinating evolved strategy for mating. Transients live in constant worry that they'll be caught red-handed with females belonging to resident males. As a strategy for coping with this risk, they retain a juvenile appearance even as they reach adulthood. Sexual selection appears to have stunted nonresident males in order to appear less threatening to the resident male, and perhaps also to the females they are trying to mate with. Studies have shown that once the resident male disappears or dies, these smallish floater males rapidly acquire the anatomy of a resident and attempt to step into his large vacated shoes.

The quandary about orangutans that has persisted for decades is why they are so antisocial. The answers have been slow in coming for exactly that reason; it's hard to compile long-term information about a species when a researcher can watch only one animal at a time over many years. But the puzzle is now more or less solved. Carel van Schaik and his students have spent years compiling ecological data on orangutans in Sumatra and Borneo, their two island homes, in hopes of making correlations with social behavior that may shed light on the solitariness question. And it turns out that orangutans, while certainly never as social as gorillas or chimpanzees, do occasionally aggregate in temporary groups of a half-dozen or more. This occurs when fruit is heavy on the trees, which happens most often in the most fruit-rich forests in which the species occurs in Sumatra. In such places at times of intense fruit abundance, it's possible to see group sociality in the species, although it won't last after the season of plenty is over.[17]

The cornucopia of a forest is linked to the climate and especially the local soils. In parts of Sumatra soggy peat soils seem to foster enough fruit tree production that the orangutan's normal territoriality is relaxed, allowing social congregations. These temporary foraging parties don't occur on the island of Borneo, where upland forests that support orangutans are typically lower in productivity. So there appears to be a direct link between fruit availability and population density and group size in orangutans. These links are known to occur widely in the primate world, so it should not be a surprise. But it came as a minor shock to many ape researchers to learn that the time-honored label of "loner" applies to the orangutan only because of fairly immediate environmental constraints, and not due to some sort of innate desire to be alone.

The two populations of orangutans on Sumatra and Borneo are these days most often considered to be separate species. This gives us two critically rare species to worry about instead of one subdivided but somewhat less endangered species. Sumatran orangutans are in terrible trouble due to unstoppable logging operations carried out with a passion by the Indonesian government. Some of these operations have been allowed to take place in national park lands, under the rarely watchful and perhaps even complicit eyes of local forestry officials.

WHAT IS AN APE?

So there you have the cast of characters. Four great apes united by a suite of traits, most of which they share with us. All the living great apes possess a rotating shoulder that allows them to hang from beneath branches to pluck food, as well as swing artfully through the forest canopy. They all lack a tail, which all other primates other than humans possess. Although many mammals have very large brains—whales for instance—apes have a brain-to-body size ratio that dwarfs that of monkeys and most other mammals. An adult male chimpanzee's brain volume is about four hundred cubic centimeters, less than a third the size of our own brains but much larger than those of other nonhuman primates.

Their cognitive abilities develop from birth much as those of human children do. Along with cognition comes language. Perhaps no issue involving the great apes has stirred passions the way that the language question has. If you raise infant chimpanzees in a nurturing home environment as though they were human children, they will develop language skills on much the same trajectory that a toddler displays. The apes will acquire a "spoken" vocabulary of hundreds of words, and will likely learn to understand many more hundreds of words spoken to them. Since apes lack the vocal apparatus—palate, tongue, larynx—of humans, they must be taught to employ some other modality to make language. As in the case of mute children, apes can learn to use sign language, or a symbol board, to communicate with humans and with each other.

The results of such language studies, which have involved all four of the great apes, are astounding. Chimpanzees who have acquired sign language speak in simple sentences, often signing spontaneously to one another in the privacy of their enclosures. They talk about their immediate surroundings, their reading materials, their offspring. Nothing very abstract, but the same stuff of which human small talk is made.

Psychologists and linguists have had a hard time getting their minds around the expansive definition of language called for by ape linguistic abilities. We have always defined language in human terms, so putting ape communication under the same umbrella makes many researchers bristle. But this is pure species-ism. When a human toddler shows the same language competence as an adult chimpanzee, linguists tend not to be impressed with the chimp because they know the child is on the verge of a language explosion while the ape will be forever mired in linguistic kindergarten. If we saw two-year-olds in a day care center communicating in the way that chimpanzees do, we would not question whether language is being used.

ONCE UPON A TIME

Once upon a time, in the tropical forests of the Old World, there lived a great many species of apes. They were both abundant and diverse, with many dozens of species inhabiting forests from the lands that are today as distantly separated as Greece, Kenya, Pakistan, and China. The first ape, *Morotopithecus*, had diverged from ancient monkey stock just over twenty million years ago. *Morotopithecus* didn't look a lot like a modern ape—it lacked the rotating shoulder so wonderfully useful for hanging under branches or undulating between trees. It also walked on the palms of its hands like a monkey or a dog rather than up on its knuckles like a proper ape. Its molar teeth, however, are those of a true ape—very similar to our own—and confirm that this primitive primate was the progenitor of the modern apes and, by extension, ourselves.[18]

During the ensuing ten million years, known by paleontologists as the Miocene period in Earth's history, the apes took off. They diversified and radiated into a variety of niches and regions. If you walked through a forest in eastern Africa or southern China fifteen million years ago, you would have looked up at apes scampering through the trees. Even in open country, apes predominated; they were as visible a part of the Miocene landscape as monkeys are in a tropical forest today. But over time, even before the hand of man wrought their destruction, the apes began a long slow decline. They were replaced by a wealth of monkeys, their lower brethren, and today those same forests of the Old World that once brimmed with apes are filled with dozens of species of monkeys.

The great apes are as distant from one another in evolutionary terms as we are from any ape. All are endangered, and at least two of them—the orangutan and bonobo—critically so. We'll consider census numbers in the next chapter, although these are in most cases woefully fuzzy and suspect. Whatever the precise number of great apes on Earth today—probably somewhere around three hundred fifty thousand comprising all four species—they are swamped by seven billion people. That's twenty thousand humans for every one living ape, and the ratio grows more lopsided every day. The threats to their continued existence are many and differ depending on which species we are talking about. We will spend much of this book looking at the threats, and also some of the possible ways to save the apes, or at least to slow the pace of this human-caused catastrophe.

Evolutionary success is not a birthright nor is it a guarantor of survival in perpetuity. Natural selection wrought the living ape species, and like all animals their time on Earth is limited by changing environments, the emergence of competing species, predators, and the like. Some species cope well in a variety of environments. Such generalists are often abundant and hang around for many millions of years. Other species lack such evolved-in versatility. Nearly all of the billions of creatures that have ever lived are now extinct, and the vast majority of ape species are just a few more members of the club. We may someday join them. But until that distant day comes, this Earth is all we have, and the four great apes will be our only extended family. Along with their very distant relatives, dolphins and elephants, they are the most socially complex creatures with whom we share our world.

Our moral obligation to protect great apes from ourselves is not solely about their intellects, their technologies, and their kinship with us, but these humanlike abilities do provide added importance to their plight and raise enormous ethical issues about their care and welfare that most societies have not yet addressed. If apes have cognitive abilities that small children or cognitively impaired adults lack, does that mean we must morally and legally accord the apes fundamental and inalienable rights?

The issue of legal rights for apes is not an academic one, nor is it unrelated to conservation and extinction discussions. Even in the community of conservationists, scientists, and activists who work to protect the last living apes, there is a vast schism in mindset. For those who hope to protect the last patches of ape habitat and thereby save apes in the wild for future centuries, the natural population, the species, and the natural habitat are the bottom line. Individual animal rights don't compare in importance to species survival. But for many activists, individual rights are what matter most. Hence the disagreement between those who believe preserving biodiversity sometimes necessitates sacrificing individuals—culling deer herds when they over-strip their resources, for instance—and those who place each deer's life on a par with the health of the population. Neither side accepts the worldview of the other, for the most part.

I cannot accept that an individual animal's life can ever be compared with the health of a population or species. In other words, species' welfare and overall biodiversity matter above all else, including the rights of any individual animal. One of the greater ironies in the conservation movement happens when animal rights advocates—who themselves are aligned on a wide spectrum from pragmatists to People for the Ethical Treatment of Animals (PETA)—campaign against hunting, the license fees for which may be the savior of wild animal populations that are being hunted. In the mind of an ecologist, the needs of the many far outweigh the needs of the few, or the one.

Great apes are in terrible trouble, from a variety of threats that we will examine in the following chapters. Being granted fundamental individual rights might be a boon for those currently living in labs and zoos. But for the great apes living in the forests where their kind has evolved over the past several million years, there's no time for and very little benefit to be gained from debating just how deserving they are of liberty and the pursuit of happiness. They need our help right now.

TWO

Homeless

The hike from the pebbly lakeshore to the top of the ridge takes a bit over an hour. It would be shorter than that—it's only a few miles—but the trail is steep and the day is hot, the dry-season sky tinged brown with dust. Tiny waves slap the beach of the great blue lake, and a hundred feet from the water's edge a trail dives into the forest. We leave the brain-poaching equatorial sun and enter a bower of thickets, gurgling streams, and fruit trees. Contrary to the popular image of a tropical forest, there are few towering, majestic trees. Along the streams the trees are large and densely grown; anywhere away from water the canopy is low, often broken, with patches of grassy clearings. There are lizards and safari ants underfoot. Coppersmiths call all around, their songs a metallic anvil-clink. The distant sound of chimpanzee pant-hoots comes faintly every so often on the afternoon breeze. The trail doesn't meander much at all. It climbs the hill slope like a mall escalator. Past the thickets of the lower hill slopes, the trail crosses a few rocky viewpoints and climbs a bare patch of grass to a stand of gnarled fruit trees. From here it re-enters the forest and passes steeply up through pleasant glades. Dead leaves and pebbles scatter at every step. Finally the trail breaks out into the open and climbs a bit farther to a perpendicular ridge that parallels the lakeshore far below. This is the rift.

We've walked three and a half miles and gained twenty-four hundred feet of elevation, arriving at the highest point in Gombe National Park, Tanzania. Everything lying before us is the home of the wild chimpanzees made famous by the work of Jane Goodall. This is the most hallowed piece of real estate in the annals of animal behavior research. The rift soars above the forest and lake, and eagles soar above the rift. This is technically a part of the edge of the Great Rift Valley of East Africa, and the enormous body of water—Lake Tanganyika—is an ancient, rain-filled crack in the Earth, a jewel in the crown of lakes that stretch up and down the western edge of the Great Rift Valley. The point we've reached on the rift offers a three-hundred-sixty-degree vantage point. Looking westward is the bowl of a valley, bisected by east–west ridges that lead to other valleys. Far below, the beach is dotted with tiny houses and a soccer field cleared from the brush. The lake, with traditional wooden pirogues carrying local fishermen, stretches off into the haze. In the wet season, the sky is clear to the far shore, which is the eastern edge of the troubled Democratic Republic of the Congo.

If we turn around and look in the other direction, eastward away from the lake, the dusty air blankets rolling green hills and khaki plains. Everywhere there are trees; nowhere is there forest. Villages, towns, and *shambas*—small-scale farms—extend to the horizon in all directions except the one behind us. Gombe is surrounded. Goodall and the Tanzanian national park service had the chance to expand Gombe decades ago, but neither she nor anyone else in those days could foresee how quickly the forest would be gobbled up, leaving Gombe an island hemmed in by the spread of farms and villages. The national park is ten miles by three-and-a-half miles. Compared to the vastness of forests in the Congo Basin or elsewhere, Gombe is minuscule, a tiny gem packed with a priceless gene pool. Everyone connected to the place worries about the effect its small size will have on its future and that of the chimpanzees.

Most of the concerns stem directly or indirectly from the tiny size of the sanctuary. A big cat needs prey and a hunting territory far larger than the size of Gombe National Park. As the forest receded, the larger antelopes were extirpated by people, the leopards and lions disappeared, and the natural food chain crumbled. Hunters removed many of the larger animals directly, reaching the deepest interior of the forest in a morning foray. People living along the edges of the forest collected fuel wood for their homes. In a small sanctuary, such cutting leads to a forest perimeter of degraded wildlife habitat.

Indirect effects of Gombe's diminutive size are also a major concern. As Gombe and thousands of forests like it have become islands, their chimpanzee populations have lost their connection to one another and become isolated gene pools. This effect of forest fragmentation is one of the most sinister and vexing of all conservation problems. It's not enough to preserve endangered animal populations. They must be preserved with connectivity. A few miles away, across a sea of grass and farms and villages, are tiny patches of forests. These satellites were once contiguous with Gombe; the chimpanzees in them are the pathetic remnants of communities that might have migrated back and forth in earlier decades. But no longer. Although there are occasional sightings of chimpanzees in the *shambas* outside the park, gene flow—the migration that maintains precious genetic diversity—has all but ended. The consequence of deforestation at Gombe is that the population of the park, estimated for many years to be about one hundred fifty but probably substantially less these days, is utterly deprived of new blood.

When a female chimpanzee reaches puberty, she begins to venture out of the community in which she was raised and visit neighboring communities. This is a behavior tactic that reflects a deeper, evolved strategy—avoid mating with your close kin. A female visits adjacent communities, where she is welcomed with open arms. She usually returns home temporarily after a few weeks, but the visits become more prolonged, and eventually she will settle into a community and become one of its resident females. This pattern is severely constrained when deforestation turns a tract of chimpanzee habitat into an island. Gombe is bordered by a mountain ridge to the east and a lake to the west, so migration could be only to the north and south even in the best of times. Now, with the other boundaries converted to agriculture and villages, a female has very limited options when the time comes to migrate. There are at least three chimpanzee communities in Gombe. The Kasakela community, the subject of much research and media attention over the past half-century, occupies the large central portion of the park. The tiny Mitumba community lives in the far north, within shouting distance of a nearby village. The Kalande community occupies a large region in the south. Little is known about this community despite decades of monitoring. So each female aspiring to emigrate has very few options and eventually may end up breeding with kin. Inbreeding will unquestionably become a health issue for the Gombe population and for many other apes stranded in tiny islets of forest in coming decades. Perhaps some future generation of conservationists will genetically manage isolated populations to introduce new genes from other forests and maintain some heterogeneity. For now, we can only protect what we have left and hope for the best.

Habitat loss is the be-all and end-all crisis facing wildlife conservationists, because it trumps all the other problems discussed in this book. As development is the way of the world, so is the cutting of forests that goes along with it. The four living great ape species have the misfortune to live only in tropical developing countries, where putting food on the table often means taking natural resources from close at hand, and where regulation and enforcement of protecting natural areas is minimal. At current rates of deforestation in the tropics, the United Nations Environment Programme estimates that chimpanzee, gorilla, and bonobo habitat in Africa will have been reduced from its current levels by ninety percent or more over the next twenty years.[1] The estimate for Asian orangutan habitat is even worse; by 2030 their forest world will be essentially gone except for a few national parks and reserves—some of which are protected only on paper.

The causes of habitat loss are many, and they are obvious. It's the solutions that are unclear and painfully difficult. If the Earth is rapidly being converted to pavement and steel in the affluent nations, it is even more rapidly becoming endless farms, villages, and towns in the impoverished, developing regions. As the wealthy nations industrialized, the cities boomed and became population centers, drawing people away from rural and remote regions. The situation in much of Africa is still the opposite; the vast majority of humanity there lives in villages and farms that stretch across thousands of square miles. Although there are also burgeoning cities scattered across the vast continent, it is the development of rural areas that constitutes the imminent threat to the survival of the great apes.

So we turn first to the destruction of the great forests that used to blanket the entire equatorial belt. The Amazon Basin in South America, the Congo Basin in Africa, and the remaining rain forests of Southeast Asia. These are the last and only strongholds of great apes on Earth. They are also the repositories of vast natural resource wealth. In the battle between biodiversity and biowealth, biowealth generally wins. There are priceless minerals and gemstones at stake. There is oil under the trees and soil. And the forest itself, the biomass of ages standing on terra firma of the rain forest, contains vast wealth in the form of timber. Visit your local lumberyard and you will see planks cut from tropical hardwoods felled in Asian forests. Some of the major retail chains—Home Depot for instance—have adopted policies aimed at eliminating the use of tropical hardwoods by purchasing lumber only from sources practicing responsible, sustainable forestry. This practice is overseen by a European monitoring body, the Forest Stewardship Council. Over the past ten years, Home Depot has decreased their use of tropical hardwoods to a point at which nearly all products (ninety-four percent according to company pamphlets), from planks to pencils, are produced from U.S.-grown trees. Home Depot literature claims they have essentially eliminated the use of tropical hardwoods from the Amazon Basin. The positive impact of such policies from a mega-chain like Home Depot is global and laudable, but the company still only accounts for a tiny fraction of the wood products sold worldwide.[2]

Few large retailers make the claims about ethical purchasing of wood that Home Depot does, with the exception of eliminating sales of nyatoh (a tree with the botanical name *Palaquium*) products. Nyatoh is an Indonesian rain forest tree often misleadingly labeled as "teak." It is widely used to make patio furniture. In fact, it is more common to find retailers bragging that they have been able to keep prices down "for the sake of consumers" (in other words, to keep sales volume high) by selling tropical hardwoods. Walmart is one such retailer that in the past has relied heavily on Indonesian rain forests for their products but has, along with Kmart, stopped selling nyatoh furniture and other rain forest timber products.[3] Unfortunately, Indonesian forests still produce most of the wood used in industrialized parts of Asia. Japan discards an estimated twenty-five billion sets of disposable wooden chopsticks every year, nearly all of which are imported from China. China churns out nearly fifty billion pairs of chopsticks annually, cut from the hearts of twenty-five million Indonesian trees. Think about those numbers the next time you dive into your dumplings or your sushi.

The logging of the world's forests is largely unsustainable. The Congo Basin is vast, but decreasing at several million acres per year. At that rate, most of the forest will be gone in fifty years, and remaining good habitat for apes in far fewer years. Most gorillas, chimpanzees, and bonobos still living wild in 2009 occupy forest that is controlled by logging concessions and is or will be slated for cutting. Fewer than one in five African apes live in protected forests. The message is plain. Implement a sustainable plan for environmentally sensitive logging across great ape habitat and reduce hunting, or all the other efforts currently underway to save the apes will be utterly meaningless.

The growth of the Chinese economy has wrought environmental disaster in many ways. China's appetite for exotic animals as food items has vacuumed all of Southeast Asia of turtles and tortoises (eaten widely), and virtually every other creature is in danger of the same fate. This happened because the Chinese had already wiped their own nation clean of many kinds of animals, which ended up in supermarkets, herbal medicine stores, and restaurants. Chinese cultural superstitions about aphrodisiacs have led to near-extinction of some species of rhinoceros (taken for their horns). China had nearly logged all of its own forests too, until the government woke up to the fact that wholesale logging can portend environmental suicide. The overall forest coverage remaining in China today, in proportion to its total land area, is about half that of most of the world's nations. When the great Yangtze River overflowed its banks in the 1990s, killing more than three thousand people and leaving large areas uninhabitable, the government blamed deforestation on the shores and banned further cutting. China then turned its attention to protecting its own tropical forests and quickly became one the world's leading wood importers. The wood comes from rain forests mainly, in Indonesia and Malaysia and from central and western Africa.

You might expect that China's worst logging impact in its insatiable quest for timber would be in Asia, but that is only partially the case. It's true that Chinese logging companies operate widely, both legally and illegally, in Indonesia. But China has been very active in cultivating relationships with African nations with the intent of expanding large-scale timber operations there. It tripled its trade relations and doubled its overall investment in Africa between 2005 and 2010.[4] On the surface, it's a win-win. China's new pledge to inject financial assistance to the poorest developing countries could be a boon in an era when a global economic slowdown has made it difficult for the United States and Europe to maintain their levels of financial aid. But the Chinese government has another plan in mind altogether. The nation is banking on Africa to feed its skyrocketing hunger for new sources of wood.

China's inroads into both the Congo Basin and the Amazon Basin could mean ecological doom for those regions because of the scale of China's timber desires. Chinese timber extraction practices are also fraught with corruption. Studies of global patterns of corruption show that China ranks behind only India in the prevalence of bribery and the like used by businessmen to negotiate their own terms of sale and export. Bribery in business practice is rife in Africa too, and Chinese companies act accordingly. Three-quarters of the wood that China extracts from African forests may be illegally gotten, according to some estimates. Chinese companies can avoid export taxes by lying about the quantity of timber they ship back home, presumably paying off officials in the process. Even if conservation measures in remote African forests are nearly impossible to enforce, quotas and restrictions on extraction do exist. Local officials are bribed to overlook extraction quotas and sustainability policies. Customs officials at the ports are bribed to allow cargo ships to leave with vast quantities of untaxed, unapproved timber bound for Chinese mills and distribution centers. Cabinet ministers and other high-ranking officials in some central and western African nations rule over kleptocracies in which politicians gain office through connections, try to stay in office long enough to amass a small fortune, and leave office only because someone more clever and more dishonest topples them for the newcomer's gain and the incumbent's loss. Making money on the side is simply part of the political process in many parts of the world, all the more so in poorer nations, of which the African continent has the world's greatest supply.

PALM OIL RULES

Indonesia is the favored playground of global logging consortiums. The world's largest paper companies, the global leaders in disposable chopsticks manufacture—they all come to Indonesia. Asia Pulp and Paper and their business partners have a plan to cut down essentially all remaining natural forest that is not in national parks in Jambi Province, Sumatra. This move would be less shocking were it not for a recent pledge by the Indonesian government to do its utmost to protect their forests of highest conservation value, which include the forests to be cleared in Jambi. The decree was announced not long before the license was granted to Asia Pulp and Paper.[5] The forests to be cut for pulp include one of the last strongholds of tigers and elephants in Sumatra, and the site of the most successful orangutan reintroduction ever accomplished. Years of work spent figuring out how to enable captive orangutans to return to the wild will be lost. And not only wildlife are at stake. There are indigenous people living in the areas to be logged. The Talang Mamak and the Orang Rimba both live in the forests of Jambi Province, relying on forest products for their daily existence. So Asia Pulp and Paper is bent on carrying out a plan not unlike those executed in the nineteenth-century in the wilds of the western United States; take the land, kill the wildlife, drive the native peoples elsewhere.

In Indonesia, palm oil is big, big business. Although the oil palm originates in Africa, where it has long been used, the Dutch introduced the tree to their Asian colonies, where it rapidly became a cottage industry for landholders. We use it in its refined liquid form, but at room temperature it's a semi-solid and is used in paste form by many villagers in developing countries. These days palm oil is produced from vast plantations on land that was recently pristine rain forest. Indonesia and Malaysia produce most of the world's palm oil, and the consumer market for the stuff has exploded by nearly fifty percent in the past twenty years. If you've eaten a Twinkie or a packaged cupcake lately, you've tapped into the palm oil mother lode. Palm oil plantations need a lot of land, and in Indonesia that land was quite recently rain forests that held many wild orangutans. Ironically, the percentage of wild orangutans living on protected land in Indonesia (about twenty-five percent) is still higher than the percentage of gorillas, chimpanzees, and bonobos living on protected land in Africa. But you'd never know it from the rate and style of deforestation in Indonesia. Although orangutans still face some pressure from hunting, it's safe to say that palm oil is, along with widespread burning of forests, by far the most pressing problem in the crisis the species faces over the coming decades, during which most, if not all, orangutans on unprotected lands will disappear. What's more, palm oil plantations are rapidly expanding, almost entirely without consideration for the forest they are replacing, or their inhabitants. Forested lands are often cleared by burning, and the burns sometimes don't limit themselves to the plantation but scorch good forest next door.

The problem is not just the inexorable bulldozing of orangutan lands to make way for plantations. In many areas of Malaysia and Indonesia, plantations form a crazy quilt that carves up forest, leaving small remnant populations of orangutans and other wildlife with no viable future. And plantations themselves are not the only culprit; the infrastructure needed to support the industry is as destructive as the bulldozers. Asia Pulp and Paper built one such road, perhaps legally or perhaps not, in order to facilitate the planned mass extraction of wood to its pulp mills. On the island of Sumatra, a road-building project in the famed Leuser ecosystem will divide two of the largest existing populations of Sumatran orangutans.[6] When a road is built, the forest is opened to all manner of invasions. Immigrants use the road to settle illegally inside the protected forest, which leads to poaching, illegal logging, and further deforestation.

This is exactly what happened in the 1980s in Gunung Leuser National Park. Once the habitat has been further degraded by the influx of humans following road-building, it becomes far easier for commercial forces to argue for projects of all sorts on the outskirts. In the end, the main economic benefits flow into the pockets of a few wealthy developers, agribusinessmen, and the government officials who granted permission for the projects. Precious little trickles down to the local people, even though it is they who are often cast as the villains in the endless cycle of forest-cutting and wildlife extinctions.

It doesn't have to be this way. Palm plantations could be grown on abandoned land that is lying fallow, already logged and now useless. The soil quality would not be ideal and the land titles for such tracts are often legally hazy. But more effective management, with a push from international conservation organizations, could help a great deal in making wasteland useful and avoiding the need for wholesale logging of pristine forests, as is the current practice. Instead, Indonesia is considering redefining the term "forest" to include oil palm plantations. This would be like redefining a city park as a national park.

There are a few points of light. Carbon payments have been suggested as a way to encourage Indonesian companies to scale back forest cutting in exchange for either carbon credits that can be traded on the international market or just in exchange for cash payments. A price per ton of emissions can be set by an international body, and logging and industrial companies can be persuaded to stop their logging in the most environmentally valuable and sensitive forests. In Sabah, a Malaysian province on the island of Borneo that is home to many orangutans, conservationists may break bread with the devil. There, palm oil industrial captains and conservation biologists are trying to craft a plan to make palm oil plantations sustainable. This sounds oxymoronic to many environmentalists. Both sides agree that there is almost no land left in Malaysia on which to plant new plantations, and conservationists certainly feel that orangutans have been pushed to the absolute brink, partly by forest clearing for these same plantations. But now the oil palm industry is agreeing to environmental assessment before new planting, and not planting at all in areas of critical conservation concern.

Palm oil produced from plantations that are managed in an environmentally conscious way is more expensive, however. Expensive products that make consumers feel good about their purchase sometimes succeed in a small way; shade coffee is one example. But palm oil has seen no such success on the environmental ethics front yet. In Europe, environmentally friendly palm oil has been marketed at a price about ten percent higher than its competitors, with very modest success. Meanwhile, companies continue to churn out tons of the product from unsustainable sources. Consumers find it nearly impossible to determine which oil products were harvested sustainably and which are the product of old-fashioned land clearing and burning. This opacity works to the great advantage of the palm oil conglomerates, who can all claim sustainability for their operations whether that is the truth, a half-truth, or a bald-faced lie. Watchdog groups have even begun to urge banning the majority of advertisements purporting to claim sustainability for palm oil products, because most have turned out to be patently false. Meanwhile the forests burn, the orangutan numbers plummet, and the government officials and industry leaders agree to more rounds of discussions about what to do.

Sumatra has lost twelve million hectares of its forest habitat since the mid-1980s.[7] This represents about half the forest remaining on the island, and the rate of destruction is increasing. One critical issue in gauging the effects of orangutan habitat loss is estimating this. Estimates are based on projections, and projections are notoriously prone to be dramatically wrong. Bad data in, bad data out. For instance, a group of conservationists led by Serge Wich recently downgraded the estimated number of Sumatran orangutans surviving in the wild from 7,500 to 6,600.[8] That's an overnight loss of twelve percent of the Sumatran orangutans on Earth. Such a precipitous drop was caused simply by updating the census and reviewing how openly data are shared and collated with other researchers. In this case, an area of western Sumatra long thought to hold orangutans in fact does not. The results suggest a very short timeline to extinction for Sumatran orangutans, which may well become the first great ape species to go extinct, perhaps as soon as 2050. Applying the same, more conservative census techniques to the much larger population of orangutans on neighboring Borneo, we can guess that the figure of about fifty thousand accepted for the island may also be high. So the overall total of wild orangutans on Earth today is likely lower than fifty thousand and shrinking rapidly.

We speak of sustainable development, a model for the future world in which wild lands and wildlife can co-exist with the competing needs of a hungry, booming population. Indeed, can there be any other goal for the future? People living near wildlife, if given access to a better quality of life, should begin to refrain from exploiting the ecosystem the way their ancestors always had. Foreign aid groups have worked diligently in some areas to assist local farmers to grow more eco-friendly crops that will also be more profitable. Tourism may be one route to sustainable development, although it won't work well in orangutan habitat in Indonesia. The sort of glamorous mega-animals that live in the open plains of East Africa just don't occur in Indonesia, and whatever large, flashy animals that do occur there tend to be shy creatures of the deep forest.

Many conservationists are, however, skeptical about the real prospects of sustainable development. Some grassroots, low-impact programs have made great strides. Other organizations, top-heavy with too many expatriate workers driving SUVs and speaking languages that are not understood by the intended recipients of the development assistance, are largely a waste of time and money.

ORANGUTANS IN HELL

Indonesia is burning. Every year during the dry season, the peat swamps and peat forests of Kalimantan and Sumatra become a smoldering furnace of vast scale. Peat swamps cover an estimated 34,000 square kilometers on Sumatra (about seven percent of its land area), more than four meters deep in some areas. Peat swamps form when well-drained land surrounds poorly drained land and waterlogged soils develop. The forests growing on top of them in Indonesia tend to be rich, highly productive, and ideal orangutan habitat, brimming with fruit trees. And because of the quantity of peat, the swamps are enormous carbon sinks—an estimated fifty million tons of carbon sit in Indonesian peat swamp forests. These forests burn with a thick smoke that blankets the landscape like a pea-soup fog. When the government encourages the drainage of peat swamps to convert forest to farmland, the result is the creation of vast areas of tinder. Not only does it burn terribly with a dense smoke, the fires may smolder for many months after the flames have gone out.

Wildfires are a natural feature of life in many parts of the world. Annual hot winds fan flames in the hills behind my southern California home, threatening lives and real estate and charring the landscape. Brush fires in Australia burn across thousands of square miles on a regular basis, sometimes coming within shouting distance of Sydney. Houses may burn and insurance companies may cry foul, but the problem lies with the real-estate developers, not with the ecosystem. These season-ally dry tinderboxes that I mention have burned periodically for millennia. Fire is truly a natural part of the landscape in many regions. Not so in Sumatra and Kalimantan. Only when people drain fertile peat swamps do they become completely unnatural kindling for human-created fires. Forests in Indonesia are not adapted to an endless cycle of fires. Trees and plants are not geared to protect themselves from fire or to spring back into life quickly afterward. So all smaller trees are killed outright, and larger trees, lacking the thick bark that trees in fire-prone regions typically evolve, are mortally wounded or left unpro-tected against infection and insect invasion.

In 1997 fires set for land clearing or perhaps as retribution in land ownership disputes started a chain of fires that burned across the island of Borneo. The fires that year produced car-bon exceeding the carbon uptake of the entire Earth's atmo-sphere and were visible from space for a full year. Well over a decade later, more than one hundred of the fires still smolder in the peat beneath the ground, bursting forth and fueling new forest fires every dry season. The 1997 fires were disastrous for land-clearing advocates as well. The fire damage cost Indonesia an estimated twenty billion U.S. dollars, most of it in lost rural income. The air pollution generated by those fires extended to Malaysia and Thailand, creating respiratory problems for mil-lions.[9] Fires as massive as those that began in Borneo in 1997 have a global reach. One study estimated that if the annual peat fires of Indonesia could be eliminated, the mandates of the

Kyoto Protocol for reducing the effects of human-caused climate change could be achieved without any further action by the nation of Indonesia.

The only viable solution to the peatland fire crisis in Indonesia is preventing the fires in the first place. This means ending the drainage of peat swamps so no new tinderboxes are created. Draining the swamps also serves to prevent wet season flooding, however, so many land owners are adamant about draining, even if they are not planning to burn forest. A decrease in burning would of course also be a great help. Unfortunately, burning is the easiest and cheapest way to clear land for agriculture. The carbon payments many international conservationists support will work only if the cost of persuading landowners and industry to reduce emissions is low enough to be sustainable, and if there is a clear link between the carbon reduction and the preservation of biodiversity in key ecosystems. The phrase "now or never" is thrown around a great deal, but in the case of orangutan habitat in Indonesia, and particularly in Sumatran peat forests, unless a system is in place soon, there will be little point in trying.

HOW LOGGING AFFECTS APES' LIVES

Any field biologist who has spent time in tropical forests in which primates live has seen firsthand the effects of logging on apes and other primates. I once planned a survey of a rare species of gibbon, the hoolock, along with other primates living in a forest in Bangladesh. Arriving at the entrance to the forest reserve, rumbling along a muddy track with exotic birds flying past and the distant call of gibbons in the air, I was excited. I stepped out of the Land Rover, pushed through some trees, and entered the main trail passing through the heart of the forest reserve. Only then did I hear the chopping. Every few seconds, from every direction came the sick sound of people hacking at trees with machetes. Every once in awhile came the crashing thud of a tree as it was felled, and brought with it to earth all the tangle of plant life that had thrived around and on it. Time after time I encountered caravans of people, usually a man carrying the murder weapon and some young boys with tree trunks and enormous branches perched on their heads. During the course of several days spent in that forest, I saw a measurable fraction of its large trees cut and removed.

All this logging activity was local—the wood was likely for fuel or building in the local villages, with perhaps only some sold commercially. And it was all of course illegal. But the large posted sign that says "logging prohibited" isn't any sort of deterrent when there is no enforcement, or at best enforcement only for those who cannot afford to pay a small bribe to the local forest rangers or police. Forest cutting like this is most common at the forest perimeter, where a buffer zone of degraded forest develops that rings a more pristine, undisturbed interior. But when the commercially valuable, preferred trees are logged from the perimeter, local people move inside the forest, and in no time, you have a national park or reserve that exists in name only, as a pathetic reminder of its intended former self.

Most of the remaining great apes do not live in national parks or nature reserves. They live in unprotected forest. A large percentage of their habitat—more than half in some regions—is leased by the local government as a logging concession to foreign, most often European, corporations. David Morgan and Crickette Sanz, both conservationists, estimate that more than a third of the most important conservation areas in western and central equatorial Africa are slated for logging.[10] The future of great apes on Earth is therefore to some significant extent going to rise or fall based on their ability to survive in logged forest. And that will depend directly on the kind of logging that is carried out.

Not all logging projects are equal. Some forests are clear-cut, as when the forest is removed entirely to plant oil palm groves or agricultural fields. Clear-cutting transforms a vibrant ecosystem humming with complexity to a barren, sun-scorched plain in a day. Most conventional logging is more selective in the sense that the logging company does not seek to fell every tree, just the commercially valuable ones. The logging company goes into the forest, trashing everything in its path to get to the large trees that it wants to extract and sell. This process is of course utterly destructive too, because it not only bulldozes wide swaths of forest to get to the desired stands of trees, it also damages soil and root systems, and often leaves behind a forest degraded beyond its ability to regenerate. Other logging is far more selective and has a far less malignant impact. Either logging companies or local people may seek out particular tree species and log them—mahogany, ebony, and other highly sought furniture woods, for example. If these are not species intensively used by the primates in the forest, then the impact of logging will be mainly on the logging site itself. Even in selective logging, people working in the forest poach wildlife, build roads and bridges, and cause other disturbances. If such road and bridge building is kept to a minimum or at least routed carefully, selective logging's impact, though long lasting, does not have to be fatal to the forest ecosystem as whole. Finally, environmentally ethical logging can be carried out, usually under the monitoring and supervision of a governing body like the Forest Stewardship Council which will certify that the logs extracted have been taken responsibly.[11] The mode of extraction will be monitored, and regulations will be in place to prohibit logging crews from hunting wildlife or carrying out other environmentally destructive activities.

No matter how ethically the logging is carried out, the forest is altered for a very long time. When large, spreading trees are felled, huge gaps in the forest canopy are created through which direct sunlight hits the forest floor, often for the first time in decades. A new ecosystem is born, with new plant species and the animals that depend on them. It will last until the light gap is filled with sapling trees, which ultimately win out and take the column of light vacated by the logging victim to nourish themselves as they grow. This process of succession happens all the time in any dense forest, but even the most selective logging accelerates the process over a wide area to the point that the forest is altered in a profound way. It also changes the composition of wildlife in the forest by removing trees that are key habitat for some species and by creating noise and disturbance that will drive away any animal with the ability to change its normal home area. Even very selective logging and the creation of roads needed to access the desired trees indirectly results in forest fragmentation. People now have access routes into the forest, leading to greater hunting pressure and to more and more road and town building. This proximity brings wildlife into even closer contact with people.

THE REAL COSTS OF LOGGING

Logging affects each animal species in the forest differently. Some animals actually benefit from logging—small antelopes, buffalo, and even elephants may find more food to eat, since the shrubby secondary growth that springs up in places where towering trees were felled is easier to digest. But primates suffer. The primatologist Colin Chapman and his colleagues reported population trends of primates living in Kibale National Park in Uganda over a twenty-eight-year period following logging. During British colonial times and for some years afterward, Kibale was logged. Some parts of the forest were logged selectively, other areas were not logged at all, and still other areas were simply clear-cut. Three decades later, the diversity of primates in the clear-cut areas had still not recovered.[12] But the densities and diversity of primates living in selectively logged forest was no different than that of primates in forest that had never been logged. From a monkey's or an ape's point of view (though not necessarily all other animal species), selective logging may be sustainable, depending on whether the tree species selected for logging are critical food resources for primates.

Conservationists agree that of all the primates, chimpanzees are harmed most by logging because they are dependent on mature fruiting trees for their preferred foods. In logged forests, densities of chimpanzees are almost invariably lower than in unlogged areas. Selective logging may not harm gorilla populations terribly; at least one study found that after initial declines in population following logging, gorillas rebounded well. Perhaps the initial stress of the disruption to the forest caused by logging gave way to benefits from the new abundance of plant growth. David Morgan and Crickette Sanz agree that the impact of logging on chimpanzees is worse than it is on gorillas. This could be because chimpanzees rely more on key fruit-tree species, or it could be that chimpanzees are more sensitive than gorillas to the influx of machinery, loggers, and attendant noise and chaos in their habitat. Perhaps because chimpanzees are intensely territorial, they regard a wholesale intrusion differently than gorillas do. Instead of just shifting their travel patterns and home range, chimpanzees may consider the destruction of a portion of their community territory to constitute irreparable harm, with high stress accompanying it.[13]

There are nonetheless chimpanzee populations that appear to be doing well after logging. Around the Dja Reserve in Cameroon, where logging has radically changed the landscape immediately outside the boundaries, chimpanzees live at densities comparable to their numbers inside the reserve. One could argue that the numbers inside the reserve have also been depressed by logging, hunting, or other human factors. In some areas a census for great apes has been taken twice within a few years and has yielded radically different results each time, making it unclear whether populations have dropped precipitously or the census techniques differed and so produced different numbers.

Could orangutans be saved by their ability to live in logged forests? The burning and wholesale clearing of forests are as lethal to orangutans as they are to all other wildlife. Until recently it was widely accepted that orangutans could not survive in forests that had been logged in any way at all. But some studies have suggested that some orangutan populations live in forests that were selectively logged in the past. Their densities are twenty to forty percent lower than in undisturbed forest. Their diets are less fruit-rich, and they spend more time traveling and less time at rest. Presumably the lower density is a result of the lack of fruit, which is the high-energy food that orangutans crave. Bornean and Sumatran orangutans also differ in their diet preferences and even in their abilities to crack open hard-shelled fruits. But the exact ecological mechanism that lowers numbers of orangutans in logged forest isn't clear. The real problem, according to Carel van Schaik, an orangutan researcher, may be that following a period of selective logging, the forest is rarely left to its own devices to recover in peace for a few decades. Instead, loggers simply move on from the original target timber species to less valuable ones until all the decent-sized trees that provide orangutans with fruit and shelter are gone.[14] The bottom line for orangutans seems to be that while no logging is far better than any logging, populations can recover to some degree from selective logging, provided that selective tree removal doesn't lead to yet more logging.

There are specialists and generalists in the animal world. Specialists are species that have evolved to be highly adapted to a particular narrow set of environmental factors—perhaps a diet or a type of vegetation without which they simply can't survive. The ivory-billed woodpecker, extinct these past seven decades except in the minds of a few ornithologists and birders who hold out hope, was doomed by its very narrow habitat requirements. The massive bird had to have mature, huge cedar trees for nesting, and they had to be located in remote swamps free of human disturbance. Once the loggers took out those valuable trees, the bird was a goner. Although there are no longer any ivory-billed woodpeckers around, other species have not only survived but thrived in the face of humanity's spread. White-tailed deer abound today in numbers that dwarf their population size at the time of European settlement of the

Americas. As the great forests of the eastern United States were replaced by a vast network of farms and towns, the deer found themselves with more to eat, not less. Every acre of forest that was felled left forest edges full of the weedy plants relished by deer. Deer also adapted extremely well to living in close proximity to people. Once hunting declined, deer populations boomed, as any driver who has had to swerve to avoid hitting a deer knows all too well. White-tailed deer are generalists. They're the mammalian equivalent of weeds; species once confined to wild places but set free by a changing environment to colonize our world as fast as we do. House sparrows, house finches, robins, raccoons, grey squirrels, and of course many species of rats and mice followed suit.

These are only a handful of generalist species that are rapidly becoming the only wildlife most people ever see. Nearly all other animals and plants suffer or simply disappear in the face of encroaching humans. The great apes are among these species. So the real question is not whether any of the apes can thrive in the wake of development, but whether they can cling to existence in any form with people and their activities moving among them in new waves every year. If we sit back and wait for ape populations to disappear from unprotected forests in the hope that wildlife sanctuaries and national parks will become their inner sancta, we may well miss the window entirely for preserving any of the ape species.

An important effect of forest fragmentation is that apes and humans are brought into much greater contact than if the apes lived only in deep forest areas far removed from towns and villages. As the forests of the world have been cut, a new field of ecology has been forced into being; the ecology of forest fragmentation. We have known about the effects of fragmentation for decades. In the late 1970s the ecologists Thomas Lovejoy and William Laurance began a now-famous ecological experiment in the Amazon basin of Brazil. At the time the Brazilian government was promoting the clear-cutting of the Amazon rain forest by offering tax breaks to land owners who logged their properties. In one region, however, the government mandated that half the forest must be preserved from logging or development. Lovejoy and Laurance proposed that the land owners leave some of their forest uncut in order to conduct an experiment on the effects of forest fragmentation. The land owners cut their forests in square patches, leaving a number of rain forest plots ranging in size from one thousand hectares (about twenty-five hundred acres) to just one hectare (about two-and-a-half acres). The ecologists monitored these patches over the following years; indeed they are still being studied to the present day.[15]

The results are as important as they are disturbing. The smaller patches lost their biological diversity quickly. An assortment of factors that affected the smaller patches far more severely than the largest patches took a rapid toll. Large trees fell at a higher rate in small patches, apparently because wind gusts were more likely to reach deep into the interior of already-reduced patches. These fallen trees left light gaps, changing the ecosystem and driving out plant-eating animals that relied on those trees. Other animals were also driven out by the rapidly transforming patch, with the result that within a few years the forest ecosystem that had existed there had collapsed, morphed into a degraded patch of forest incapable of supporting the biodiversity native to the place.[16] Similar studies of fragments conducted elsewhere in the world confirmed that forest patch size is strongly associated with ecological changes that can preserve or destroy the community within.

Where does fragmentation leave apes? It is always lethal. The animals have two options; move to another area or withstand the changes wrought by encroaching cleared land. Moving is rarely an option. Chimpanzees are intensely territorial, and a newcomer would face lethal aggression from residents. Instead, great apes hang on as the trees fall around them. If they survive in smaller forest blocks, they are brought into far more contact with people and villages than is good for them.

Much of the conflict that humans have with chimpanzees and gorillas occurs because of crop raiding. Far more than you would suspect, great apes surrounded by humanity end up spending a lot of their time in farmland, which is also where the farmers are. Chimpanzees are highly adaptable when it comes to human foods. They learn that wading into a sugar cane field is far easier than traveling miles to find ripe fruit. Much of the damage done to cane fields in Africa by "agricultural pests" is by wild chimpanzees, who augment their natural diets by taking domestic crops when forest foods are less available. Mangoes, papayas, avocadoes, cane sugar, rice, bananas, yams, maize, cassava, cacao, oil palm fruits, and pineapples are all relished. And, of course, the farmers hate them for it.

A farmer in Uganda rounds a corner in a trail and comes face to face with a four-hundred-pound silverback male gorilla. The normally timid gorilla reacts in panic by biting a chunk of flesh from the rear of the farmer as he turns to flee. Not far away, a sugar cane plantation is raided regularly by the local chimpanzees, which have come to regard the fields newly abutting their forest home as a supermarket. Chimpanzees in Ivory Coast leave the forest to enter cacao plantations that encircle their shrinking habitat. The farmer responds by shooting them. And back in Uganda, a chimpanzee struts into a village, enters a small house, and emerges dragging a screaming human baby. The baby's siblings give chase but to no avail. The baby is found in the forest hours later, partially devoured. A chimpanzee, apparently the same one, makes similar village raids over subsequent months, killing or maiming small children. Chimpanzees are not bloodthirsty man-eaters. But they are omnivores who love meat when they can get it. Chimps that have lived in close proximity to people have had the chance to observe how defenseless humans are, especially young humans, and simply wait for their opportunity to strike.

Crop raiding earns chimpanzees and gorillas a bad name among African villagers. Preying on humans earns them hatred and fear, and unwarranted blame. But all these attractive nuisances—food crops or the darker draws of civilization—are just irresistible temptations placed in the diminished world of great apes living side by side with people. Forest fragmentation inevitably leads to all the other critical problems I detail in this book: disease transmission, poaching, and direct ape-human conflicts.

Ironically, local people often prefer to plant and use tree species that are not native to their local forest. When given an affordable opportunity, they will opt not to cut the native trees. When I began construction of my research camp in the Impenetrable Forest in Uganda, I made an agreement with the Ugandan Wildlife Authority (the Ugandan version of the U.S. National Park Service) that we would use local materials that would not be permanent and that we would of course not cut any local indigenous trees. We built our walls of mud and wood rather than brick or cement blocks, and we used the most abundant wood available, eucalyptus. Eucalyptus, the iconic gum trees of Australian landscapes, are valued all over the world for their hardiness, low price, and rapid growth to marketable size. In our case, there was a local farmer who had turned over many acres of his hilliest land to eucalyptus farming. My doctoral student, John Bosco Nkurunungi, negotiated the deal. We looked at the trees, haggled over a price, reached an agreement, and confirmed a delivery date. At one point the owner of the grove admitted that the actual trees we were buying were not *these* trees, but rather another, more distant grove beyond the next set of hills. In any case the deal was done and not long after the local version of Fedex arrived. A human chain of small children and their mothers, each balancing a ten- to fifteen-foot tree trunk a foot thick, strode into camp and deposited their load. I had never felt so colonial in my life. When I implored my field assistants (three of them, all twenty-something men) to stop this misuse of women and children, the men laughed at the fact that I had assumed that men would do the heavy lifting. Construction was important, but delivery of the raw materials was menial, went their argument. In any case, we got our wood.

My point is twofold. First, using nonnative species can be a boon to the natural habitat of apes, depending on how and where the trees are grown. If, as is the case in many areas, natural forest is felled to create space for more lucrative trees to be planted artificially, the wildlife loses. But if local people strategically plant fast-growing trees around their villages, knowing that within a few years they will reap a profit from the trees' sale or use, then in all likelihood many naturally grown forest trees will be saved. That is sustainability. Second, it is almost always the women who interact with the environment most directly. They go out to collect fuel wood to keep the home fires burning. These trips become longer and longer as the land around the village is deforested and if no alternative trees are planted. In some parts of Africa women spend up to five hours a day simply searching for wood to burn. Working with local women has become a key piece of any sustainable development project. Helping them with new, more efficient ways to cook not only makes their lives easier but also reduces pressure on wood in an increasingly wood-scarce world.

GORILLAS AND YOUR CELL PHONE

You've probably never heard of columbite-tantalite. But you no doubt own some, and you may even have it in your pocket as you read this. Its nickname is Coltan, and it is a combination of mineral ores widely used in everything from cell phones and pagers to DVD players because of its ability to store electrical charges without overheating. When Sony PlayStation 2 was released a number of years ago, the global price of Coltan sky-rocketed overnight. That's how much the economies of the West can influence other, seemingly disconnected, parts of the world. Eighty percent of the world's Coltan supply comes from the rain forests of the eastern Democratic Republic of the Congo, where a gold rush of sorts has taken place. The profit to be made from mining the mineral is a draw that has brought thousands of immigrants to the Kahuzi-Biega region of the Congo. Kahuzi-Biega is also the home of many of the world's remaining eastern lowland gorillas. Coltan is mined in streams, much the way gold was panned in the mid-nineteenth century in California. Miners, working in awful conditions for minimal pay, sift through stream bottoms, hoping to find quantities of the powdery material which, when dried, becomes nearly as valuable as gold. In doing so, they trash the stream and level the forest surrounding it.

The social problems caused by Coltan extend well beyond the mining itself. The miners need to eat, so they hunt every living thing. Prostitutes arrive, as do others hoping to capitalize on a captive population of impoverished men who now have a little spending money. And there is big business too. Just as conflict diamonds fuel civil wars elsewhere in Africa, Coltan is a conflict mineral in the Democratic Republic of the Congo. When ethnic violence in Rwanda spilled over into the DRC and began to threaten Uganda too, the Ugandan army invaded eastern DRC. They invaded and occupied in the late 1990s to restore peace as part of a multinational effort. But they stayed for the Coltan, as well as gold, diamonds, cobalt, and timber. The regional warring in eastern Congo may have killed over five million people over the past decade. Although Rwanda, Uganda, Zimbabwe, and other nations claimed they acted in the interest of their own security and regional stability, the rape of the natural resources of the eastern Congo was a side benefit for all factions involved. In fact, reports suggest that the foreign armies operating in Congo may be more concerned with maintaining control over natural resources than carrying out their original operation. The opportunity to pay their operational expenses as well as line their pockets led military leaders operating in the lawless Congo to smuggle natural resources across the border. From there, the Coltan likely ended up in a Chinese factory, and eventually in your Blackberry.

So you and I have subsidized the rape of the land the gorillas of eastern Congo need to survive. We may not like to think that we have contributed to the loss of their habitat. But that is the way it works with habitat loss. The domino effect, although byzantine and indirect, is real, and it often starts with the thirst of the West for the products of the developing world. The burgeoning populations of many impoverished countries place intense pressure on the land on which tropical forests currently grow. But Africa is not overpopulated in the way that China or India is. Some countries on the African continent are downright sparsely inhabited, and the massive mortality of the HIV pandemic has played a role in this. Other parts of the world inhabited by apes, such as Indonesia, face environmental destruction and forest loss orders of magnitude greater than the same issues in Africa today. And of all the problems that endangered animals face, habitat loss is the one that is not going to go away.

THREE

Bushmeat

A billowing cloud of red dust follows the car as it roars into the little marketplace. The SUV screeches to a halt and a man steps out. He is well dressed compared to the villagers he walks among, looking like a city guy lost in the countryside. He walks into the open-air bazaar, past dozens of tiny stalls. Smiling shop owners call out to him as he passes, urging him to buy everything from plastic baskets to stacks of tomatoes to sides of freshly butchered beef. He ignores them and heads straight for one of the narrow back alleys. It is dank and smelly, and has a trickle of open sewage on the ground. Following it, he arrives at a kiosk where a man is slicing up cuts of beef. Seeing a familiar customer, the butcher nods to a woman at the counter, who slips into a back room and emerges a minute later with a large wicker basket draped with a blood-stained cloth. The customer peers inside, then takes off the cloth to inspect his purchase.

He reaches into the basket and pulls out what at first glance looks to be a large but withered human hand, black and twisted with the look of a mummified skeleton. There are three more appendages just like it in the basket. They are the hands of at least two gorillas, smoked and charred. The asking price for them is about five times what the customer would be asked to pay for beef, but he is okay with that. He slips some cash to the vendor and threads his way back to the waiting SUV with a delicacy to impress those at the dinner party he is throwing the next night for some important business colleagues. Meanwhile, the butcher sends out the word: he is in need of more gorilla meat and has potential new customers lined up already.

THE ANCIENT BUSHMEAT TRADE

For thousands of years, people in many parts of Africa have been hunting and eating the great apes. It's been a part of their culture and a way to put desperately needed protein on the table for their families. And though it may seem gruesome to kill and eat a creature so nearly human as a chimpanzee or gorilla, we humans eat all manner of odd foods. Westerners find it repulsive that some Asian peoples relish dog meat, even if the dogs are raised and cared for humanely until their slaughter. Meanwhile we turn a blind eye to the way we rear our food animals in factories, in the most appalling conditions, in order to have chicken on our plates. Humans love to eat lobsters, crabs, spiders, scorpions, and cockroaches—all creatures of the same ilk more or less, even though a lobster-loving American would be nauseated at the thought of munching on a tarantula. When it comes to food, cultural relativity rules.

The term *bushmeat* draws blank stares when mentioned in the company of otherwise well-informed people. Even environmentally savvy folks have little idea what it refers to, or that it is a crisis of epic proportions in Africa. But most Americans know all about bushmeat; they just don't use that term. Hunting deer, elk, and everything else that moved was part of life for a majority of Americans during the first two centuries of settlement of North America. Today millions of game animals are shot each year, and some are lovingly prepared and eaten. Americans no longer need venison or rabbit for nutrition; they are recreational food items killed and eaten for the dubious pleasure of the hunt and the kill. The situation is very different in the developing world, where saving a few pennies a day by hunting instead of buying one's meat is a necessity. Even the cost of a few bullets is a financial hardship to the average family man searching for a little animal protein and animal fat to put on the table that evening.

Were great apes being hunted and snared in Africa today solely to feed local families, conservationists would be hard-pressed to argue persuasively that the practice should be outlawed. But bushmeat, so long a staple of everyday village life, has gone commercial, and that is the problem. The upwardly mobile African who sweeps into the market in search of ape meat is a cog in the engine which drives a global black market economy that runs on the carcasses of wild animals. The thing is, many animal species can withstand human hunting because they reproduce early and often. Many antelope, deer, and other mammals can both tolerate substantial hunting levels without serious harm to their overall population and also rebound quickly from severe drops in population due to intensive hunting. Great apes, with their glacially slow reproductive rate, can neither withstand much hunting nor recover from intense depredations. Humans have been natural predators of great apes for a very long time, albeit when the human population in Africa was a tiny fraction of its size today. The bushmeat trade became a crisis only in recent years when the hunting a man might do for his family was exponentially increased by the rise of the commercial bushmeat market. What used to be shot and carried home a few miles to a village hut is now being shipped to cities hundreds of miles away. Occasionally bushmeat ends up in cities across the globe. It is a multi-billion-dollar industry, not a rural way of life.

Bushmeat is a term co-opted from the French *viande de brousse*. It refers to the meat of a wild animal when that animal is killed for food rather than sport. The developing world is set apart from the industrialized developed world only because hunting in poor nations is for subsistence rather than sport. The taste for some animals is driving them into extinction, and the hunting is poorly regulated. Only a tiny fraction of the bushmeat trade in Africa is great apes, and a somewhat larger but still small slice is all primates. Mostly bushmeat is antelopes, pigs, porcupines, and myriad other mammals and birds. And when we speak of the African great ape bushmeat crisis, we are really only speaking about certain regions of Africa. With few exceptions, people in eastern Africa do not eat primates because they consider them too similar to humans. There is little or no bushmeat trade for orangutan meat in Asia (although the trade in live orangutans there is still brisk and threatening). The East African ape-eating taboo unfortunately does not extend to central or western Africa, and that is where the vast majority of gorillas, chimpanzees, and bonobos live, and where the bushmeat trade thrives. Although central Africa still has huge tracts of unlogged and fairly undisturbed forest that look pristine, these places have been emptied of most of their large mammals, including apes. This is the impact of the bushmeat trade.

HOW IT USED TO WORK

For millennia a man, or perhaps a party of men, went out hunting with spears, hoping to bag a meal. In the old days and under the old ways, there was grave risk involved in hunting a gorilla. Gorilla meat, which tastes more like buffalo and other beef than does the meat of other primates, is highly valued not only for its taste but for the massive bounty that awaits a successful hunter. Hunters had to approach with great stealth, then face the onslaught of a furious silverback while armed only with their spears. Hunting an ape who was high in the trees was difficult work. And killing an ape who was enormously strong and dangerous when threatened or injured was simply terrifying and treacherous. They had to stalk the animals all day long, approaching to within very close range before hurling their weapons. All the while they were within charging distance of the enraged silverback.

All that changed with the introduction of shotguns, which have recently replaced the spear or bow and arrow in many regions. Now hunters can avoid much of the risk by shooting the silverback. They still have to approach with stealth, but they can shoot a female from a safer distance, then flee a hundred yards with the silverback in pursuit. He will give up the chase fairly quickly, and if the hunters hide themselves in the bushes for a hour or two, the gorillas will move on, leaving the female's carcass for the hunters to find and butcher. Even this hunting method can be perilous. The average villager is armed only with an old, unreliable, and low-powered shotgun that might not stop a charging gorilla. So the old ways still apply to some extent, and the ability of local people to severely impact a population of gorillas is thankfully limited by the feeble guns available. That, however, is changing, as I will discuss later.

If the hunter is successful and he manages to avoid the onslaught of the silverback male, or if his quarry is chimpanzees or bonobos, he is still confronted with the task of butchering the corpse. If he is close to his settlement, this may not be an enormous problem—he can return to his home and bring back men to help with the task. But if he is many miles from home, carrying the meat to his village and his family's table requires some quick action. In the case of the massive corpse of an adult gorilla, he is faced with a hundred and twenty kilos or so. It is freshly killed but will quickly begin to decompose in the tropical heat and humidity. He is armed only with a *panga* (the African version of a machete) and whatever ropes and baskets he can fashion from the natural fibers of the forest. If he has the help of other villagers, the entire corpse can be lashed to a wooden frame and carried on their shoulders to the village. But if he is alone, he must cut the body into chunks of a size that can be carried in his homemade fiber basket or rucksack.

He eviscerates the corpse of its soft tissue, no small job given the length and bulk of the gorilla's digestive tract. This offal can be cooked and eaten in the field camp or transported back quickly, as it will not keep long before spoiling. The hands, feet, arms, and legs of the animal are cut off—these are prime items to be smoked over an open smoldering fire of leaves and debris and either consumed as delicacies or bartered locally. Given the value of gorilla meat, they will probably be smoked to preserve them, stored for a time, or bartered for other necessities. The remaining torso—the largest and heaviest part of the gorilla—is then butchered into pieces that can be strapped into the fiber rucksack, which itself is carried either with a tumpline around the forehead or straps over the shoulders. Finally the head is either saved or discarded, depending on how far it must be carried and whether there is a chance to save or eat its main delicacy, the brain, before it begins to rot.

Although spears and arrows were the time-honored way to

kill large-bodied primates like gorillas or chimpanzees, African villagers have always done a lot of trapping and snaring too. A trap could be as simple as a pit whose open top is concealed with brush. Hoping that a large animal will fall into a pit is about as primitive as it gets as a way to catch dinner. Using hunting nets made of fiber from the forest is as ancient as the people themselves. Setting snares is also an ancient way that humans prey on small game. A loop is made of natural fibers, tethered to a branch overhead, and placed under leaf litter on the forest floor. When the unsuspecting pig or antelope steps in the loop, the snare is triggered and, in its struggles, the prey becomes more ensnared. The yield is low, so the number of snares set on the floor of many forests in Africa is enormous. Gorillas, chimpanzees, and bonobos occasionally step in such traps, but their nimble fingers give them an advantage over other victims, and they are often able to free themselves. At worst, they might drag a fiber snare around on their wrists or ankles for a few weeks until it rotted enough for the animal to dislodge it. That all changed with the introduction of metal snares.

An ancient hunter-gatherer would then pack the gorilla back to his village or to some gathering of his people. At this point unwritten rules of sharing came into play, which were rigidly enforced by a social code of ethics. Foraging people tended to be egalitarian; there was rarely a headman and if a man attempted to gain political power or influence over his group, he was scorned and even ostracized for it. Humility, altruism, and cooperation were the orders of the day. In fact, the hunter whose arrow or spear brought down the game might not willingly admit to his trophy. Instead, he would allow another member of the hunting party to dole out the meat, and his own family might receive less bounty from his own kill than they would have if another hunter had been the captor. These rules, nearly universal among hunter-gatherer people, appeared to serve as risk-reduction; if you succeed, you share widely so that when the day comes that you fail, others will share openly with you.

In more recent times, people settled themselves into permanent or semi-nomadic villages dependent on small-scale agriculture. A chief system developed. Even then there were rules about when, where, and how much to hunt. Villages controlled tracts of forest and made agreements with neighbors about who could hunt there. There's a time-honored myth that traditional subsistence hunters conserve their natural resources by selecting only the old, the sick, and the young prey to hunt, leaving the breeding population intact. This variation on the noble savage fallacy is belied by the actual pattern of hunting that anthropologists have observed. Michael Alvard showed that among two Amazonian tribes, the Machiguenga and the Piro, the selection of prey did not support the idea that they practiced sustainable resource harvesting. Instead, it's likely that the lack of high-powered weapons combined with the historically low population density at which subsistence foragers live contribute to avoiding depletion of prey populations. Primates, who live long, slow lives compared to, say, an antelope or wild pig, are particularly vulnerable to overhunting. But instead of rationing hunting of apes in order to preserve biodiversity, local people are more likely to kill whatever they can kill, and their predatory impact is lessened only by their low numbers and lack of guns.

It's impossible to know the impact of hunting on wildlife populations in the twentieth century before widespread commercialization. But conservationists estimate that people were consuming hundreds of thousands of tons of wild meat in each region of western and central Africa even decades ago. Unlike other parts of the Old World, there were no African wild hoofed animals very well suited to domestication as livestock. While early Europeans were domesticating wild cattle, horses, and goats, Africans were discovering that zebras and eland were not good candidates to be selectively bred into compliant sources of meat. The rain forest is also not a good place to establish pastureland for grazing animals, and much of tropical Africa is riddled with livestock diseases that make even small-scale ranching difficult. So nearly all animal protein has traditionally come from the forest. It still does to a large extent. The critical difference is the new role of bushmeat in a regional and global marketplace.

HOW IT WORKS TODAY

Times have changed. Although Africa is still a largely rural place, there are also teeming cities, along with thousands of towns that swell daily with immigrants from far-flung villages. Today, apes and other primates are hunted relentlessly in the forests of west and central Africa to supply a market demand that caters to city people who have cash to spend. In the old ways, bushmeat just meant dinner. In the modern world, bushmeat has taken on a new meaning. It is often illegal hunting or snaring of species that are endangered or threatened with extinction. It often involves trespassing into protected sanctuaries or national parks. Most important, even if the practice of eating wild animals is ancient, in the modern world it is unsustainable, especially when it involves animals that reproduce as slowly as great apes. The industry today is made possible by a logging industry that builds the roads on which ape meat is transported from forest to town. It is also facilitated by the lack of any effective law enforcement in many areas and by the lack of other sources of animal protein that might replace apes. Underlying these key factors is a cultural taste for the meat of great apes.

Bushmeat has become a key resource, with a commercial pipeline that threatens to depopulate forests that may still look outwardly pristine. It works like this. Logging concessions are granted to large lumber extraction companies in the Congo Basin rain forests of central and western Africa. These leases are often controlled by European corporate interests and sometimes subsidized by governments in France or Germany. The prime timber is deep within vast tracts of forests, far beyond the range of extraction machinery. So the company builds a road. Now there is an artery that runs miles through the heart of what was once pristine rain forest. The road enables logging trucks to penetrate to where the prized trees grow, and it allows those trees to be carried out quickly to the lumber mills. Now logging crews are being transported many miles deep into gorilla and chimpanzee habitat. Instead of also transporting food and supplies for the logging crews, the companies often just issue shotguns, with the assumption that the crews can feed themselves. And they do, shooting every antelope, pig, monkey, bird, and ape in their gun sights. But it's one thing to feed oneself, and another to realize there is money to be made. Instead of limiting themselves to bushmeat for the camp, the workers hunt or employ local villagers as hunters, and then ship the meat out of the forest on the logging trucks that carry trees to market.

The evolution of the hunting weapons available to rural people has been deadly to the great apes. When you use a gun instead of a spear to attack a gorilla, you raise the ruthless efficiency of the hunt by orders of magnitude while dramatically reducing the risk of mortal injury to yourself from a charging silverback. But a simple single-shot bolt-action rifle won't do against a male gorilla. It will critically injure, but it may not stop him in his tracks before he reaches you. So unless the hunter has a rifle with tremendous firepower, his ability to hunt gorillas is severely constrained. The introduction of guns with multiple-shot cartridges—*chevrotine* in French—changes all that. Such guns allow the hunter to discharge a wide scatter of shot into the silverback and tip the equation strongly in favor of the predatory human. Once the silverback is taken care of, the hunters can have their way with the females and their offspring. The cartridges used in such powerful shotguns have been widely available in central western Africa. Almost all are manufactured by French-owned companies based in Africa. One such factory in Pointe-Noire, Republic of the Congo, was churning out ten million rounds per year, nearly all of which were destined for the bushmeat trade.[1] To kill a gorilla, you'd buy the nine-ball *chevrotine* cartridge. And instead of thinking every hunt might be your last, you knew you had an excellent chance of success.

So now there is a business model. Once the supply and the pipeline are set up, middlemen enter the picture. They agree to buy bushmeat—indeed, they place orders—and then sell it to meat vendors in towns and eventually cities. Flights out of backwater airports have crates and bags of smoked or fresh bloody meat bound for the urban markets of Yaounde, Brazzaville, or Kinshasa. Some have reported great ape meat on the menu, clandestinely, in European restaurants. Although there are crackdowns whenever a restaurant is exposed, it's a decent bet that on the day you read this, someone in a restaurant serving Congolese cuisine in Brussels or Paris is ordering monkey or ape. It's pricey and it's illegal, as delicacies often are. As part of the cultural legacy and culinary heritage of many west and central African ethnic groups, it is an acquired taste not easily extinguished. Chimpanzee meat has been offered for sale in England as well. A recent study estimated that over two hundred and seventy tons of bushmeat passed through Charles de Gaulle airport in Paris each year. That's more than five tons per week, a total that included chimpanzee meat, which sells for about twenty dollars per pound on the black market.[2] The true extent to which such contraband makes it across the Atlantic via smugglers to be sold in the United States is unknown but believed to be substantial in the large African expatriate communities in New York City and elsewhere. When a monkey that sells for a few dollars in an African market can bring two hundred dollars on the global black market, there will always be a waiting list of smugglers eager to attempt to transport them to Europe and North America.

All along the logging road, tiny communities that once relied on the forest only to put protein on the table quickly see the possibilities for enhancing their income by offering wild game to the middlemen and the logging crews as they come and go. People who once would have lived deep within the forest in an area plentiful with game move to the road edges to become professional hunters, seeking a rapidly shrinking resource base for maximum short-term gain. Bushmeat hunting is more intense along newly constructed logging roads than it is in roadless areas, for obvious commercial reasons. And although some local cultures in Africa have ancient taboos against eating primates, these tend to break down as the overall level of hunting increases and primate meat becomes more valued.

Snaring is also an effective way to kill great apes, although they are rarely the intended targets. Chimpanzees who step into snares, even ones that may have been set for other animals, in the twenty-first century face a far greater challenge than they would have a few decades ago. Instead of a natural rope fiber snare that can be removed or counted upon to decompose, modern snares are made of ruthlessly efficient metal wire. Razor wire pillaged from home security systems may be employed. When the ape steps into such a noose, neither manual dexterity nor the passage of time is enough to remove it. On the contrary, the metal wire works its way into the flesh of the ankle or wrist, and if the ensuing infection doesn't kill the victim, it results in the loss of a limb. Such amputees now make up a substantial portion of great ape populations all over Africa. In Budongo Forest of western Uganda, a full third of the chimpanzees are either amputees or otherwise mutilated by wire snares.[3]

Now we have a resource base opened up to commerce, a conduit for bringing the resource to the open market, and many people eager to capitalize on the business model. Unfortunately, there is no silver lining for local people. Local hunters are paid very little for the sale of their catch, though it increases exponentially in price by the time it reaches Brazzaville, Kinshasa, or Brussels. Local indigenous people suffer from the influx of logging crews and their families, and of hunters from the outside, who bring disease and also a cash-based ethic that leaves the people of the forest in the dust financially.

As people migrate, they take their culinary preferences with them. Areas of Africa that were safe for great apes, at least from a bushmeat perspective, are no longer protected because of the immigration of people who arrive with a taste for chimpanzee or gorilla meat. For centuries the eastern shore of Lake Tanganyika was populated by the Waha tribe, whose culinary traditions did not include eating primates. But in recent decades, refugees from Burundi to the north and the Democratic Republic of the Congo to the west have poured into the area, and these immigrants eat monkeys, apes, and pretty much everything else that moves. One day while hiking in the higher reaches of the hills in Gombe National Park, Tanzania, I saw a scattered band of men in a valley below. They appeared to be goat herders, walking with their livestock and carrying walking sticks. But a long look through binoculars showed a very different scene. They were hunters, each armed with a long wooden spear, and what I had thought were goats were actually hunting dogs. They had come over the high rift that separates the national park and the Lake Tanganyika lakeshore on the west from the patchwork of farms and villages to the east. Although anyone in that region might have been willing to sneak into a remote corner of the wildlife sanctuary to hunt, I knew that the area beyond the rift was inhabited by immigrants from Burundi, who include monkeys or chimpanzees in their menu. And as I quickly learned, a party of chimpanzees was ambling very close to the hunting party. The dogs were sniffing excitedly at the trees in which the chimps sat. I began to holler at the top of my lungs. The distance was too great for the poachers to see me, but they heard me calling and turned to hike back to their villages. Such encounters with poachers were once unheard of in Gombe, but the influx of outsiders has led to strong suspicions that Gombe's wildlife, including its famous chimpanzees, may sometimes be victims of illegal hunting.

THE IMPACT

If you're an urban African and you want to serve bushmeat at a dinner party, you'd better have plenty of cash to spend—bushmeat can cost several times the price of cattle beef in the same market. The impact, while difficult to accurately measure, is devastating to African apes. One study of a single large market in Brazzaville in the Republic of the Congo reported fifteen thousand carcasses offered for sale in a single year.[4] Most of these would be animals other than primates, but there were nearly three hundred chimpanzees among the slaughtered animals. In Yaounde, the capital of Cameroon, up to a metric ton of bushmeat is unloaded from trains entering the city every day, headed for the markets.

Although that sounds like a staggering quantity of wild animals removed every day from their ecosystems, we don't really know how to assess the impact. It certainly sounds unsustainable—but is it? In his masterful book on the bushmeat trade, *Eating Apes*, Dale Peterson presents figures for the likely maximum sustainable in bushmeat from African forests. He notes scientific studies which suggest that, based on surveys of wildlife in African forests, such forests can sustain no more than about three hundred pounds of wild meat per square kilometer in a single year. If, Peterson reasons, about sixty-five percent of an animal's weight is edible (excluding its skeleton), then these forests can produce about two hundred pounds of bush-meat per square kilometer per year, or only a half-pound or so per day.[5]

David Wilkie and Julia Carpenter conducted a more de-tailed analysis of bushmeat's impact on wildlife populations. They found that, not unexpectedly, the growth of the commer-cial bushmeat trade is driven by urban rather than local demand and has generated a market that has depopulated areas near vil-lages of wildlife. They estimate that bushmeat consumption by households in the Congo region alone may exceed a total of one million metric tons. In the Democratic Republic of the Congo, the bushmeat biomass consumed may be nearly one ton per square kilometer of forest per year. Much of this enor-mous total is elephants, antelopes, rodents, pigs, and other mammals. Duiker antelope, small and abundant, are the form of bushmeat eaten most often. Even though they reproduce quickly compared to any primate, the bushmeat trade has sent even this animal's wild population into a tailspin.[6]

If only a portion of this total is primates—somewhere be-tween five and forty percent depending on the locale—it still represents a holocaust for the ape populations of equatorial Af-rican forests. Such consumption might have been sustainable in a time of very low human population density, especially when traditional foraging people lived mainly on plants, roots, and such. But with an increase in population, it became utterly unsustainable. This, along with the cultural value placed on bush-meat in African culture, may account for the sky-high prices for which bushmeat is sold in urban markets. For those who think that the bushmeat trade is not coming to their hometowns, occasional busts of smuggling operations tell us otherwise.

The role that primates can play in the bushmeat trade and still maintain a viable population is determined by hunting intensity and also the species' intrinsic reproductive rate. Local people could remove a huge percentage of the rodents from a forest every year without sending them into extinction because rodents have the ability to increase their numbers exponentially with rapid reproduction and large litters. The maximum sustainable yields for larger mammals are much, much lower. Peterson estimates that hunters could take only between four and twenty-five percent of the local population of an ungulate like an antelope every year before the population would decline severely. His estimated figure for the killing of primates is far less, perhaps a maximum of four percent for forest monkeys. For great apes, the sustainable yield is lower still, perhaps one to three percent of the population removed annually.[7] One recent estimate of the actual harvest of chimpanzees and gorillas in the Republic of the Congo showed five to seven percent of those species taken out of the population per year.[8] For species in which females give birth only once every four years, that is not sustainable.

In Central and West Africa today, the appetite for bush-meat is unsustainable even for rapidly reproducing small mammals, let alone primates. In the northern Democratic Republic of the Congo, Pearce and Amman estimated that between four hundred and six hundred gorillas were being killed each year in the mid-1990s. The slaughter was carried out not only for bushmeat but for infants to be sold alive to expatriates, and for the use of gorilla parts in witchcraft.[9] If these numbers are typical for all regions of the DRC, it represents a staggering toll on the gorilla population. In general, estimates of the impact are few and far between. The best we can do is tally the percentage of total bushmeat that consists of great apes. Because apes are large-bodied, they compose a relatively large percentage of the total bushmeat biomass—each gorilla carries the meat of around twenty duiker antelopes. A recent study by J. R. Poulsen and colleagues found that the advent of industrial logging in northern Congo led directly to an enormous increase in both local human population growth and a concomitant increase in the supply of bushmeat in the region.[10]

No one has done more to bring the bushmeat crisis to light than Karl Ammann. Ammann is a Swiss-born businessman turned photojournalist and investigative reporter, whose hard-headed, take-no-prisoners approach to learning about and publicizing the bushmeat trade has brought him accolades from western conservationists and scorn and threats from logging companies and some African officials. If you are familiar with the term *bushmeat* itself, you probably owe your awareness to Ammann or to the many articles that have provided accounts of his work. According to Ammann's investigative journalism, populations of great apes that are not protected within sanctuaries and live within twenty miles of a logging road or a town are being devastated by hunting.[11] Ammann is careful not to blame the local populace for this mass slaughter. It is driven, in his view, mainly by the timber industry's relentless quest for new logging concessions and new ways to open up existing forests that remove the last protective barrier—remoteness—between apes and their human hunters.

Ammann's investigative style of reporting has not won him many friends in African government or commercial circles, but his willingness to push the bushmeat issue has appealed to hardcore conservationists. Mainstream conservation groups are another matter. Ammann believes that most conservation agencies spin a message to the public about a crisis with a possible resolution. Donate now and your world shall be saved. Ammann believes this is nonsense; without radical action apes are going to disappear, and soon. He sees the approach of most western conservation groups toward the bushmeat crisis to be inconsistent and ineffective. He focuses on combating the illegal meat trade by increasing awareness, both in the West and in bushmeat-consuming regions. This approach means discarding to some extent all respect for time-honored cultural practices in African cultures, and this has riled the more politically correct, anthropologically sensitive factions in the conservation movement. Ammann responds that cultural sensitivity loses its place when the modern bushmeat poachers wear camouflage and use high-powered rifles. This is mass murder in the name of commerce, not the traditional ways of an indigenous people. Besides, the population boom in bushmeat-rich regions is such that the illegal harvest is unsustainable for more than a few years. Passing judgment on what people choose to eat, and doing so loudly and critically, is hardly a diplomatic approach. But then few people have ever characterized Ammann as a diplomatic person. Ammann might respond by pointing out that the bushmeat crisis is just that—the most severe conservation crisis facing the great apes—and therefore one warranting bold measures.

Ammann prepared glossy magazine articles with lurid photos of the severed bloody heads and limbs of gorillas and chimpanzees, along with shots of local African people preparing to carve up apes or loading dismembered carcasses onto smoking racks. These photographs and the accompanying articles were so graphic that many North American magazines initially refused to publish them, which drew further ire from Ammann. When they were ultimately published, they offended, embarrassed, and provoked African leaders. Some of these leaders had been more than willing to host conservation conferences. At some of these, bushmeat was actually considered for the banquet menus. This sort of cognitive disconnect is what Ammann has succeeded in revealing through his journalism.

Ammann's combative style has not prevented him from obtaining the support of the European Union, which enacted a ban (temporarily, at least) on the manufacture of multiple-shot large-ball *chevrotine* ammunition. He has at times served in advisory roles and as a spokesman for global animal protection societies. So progress has been made. But whether this progress is just a drop in the bushmeat bucket or the beginning of a change for the better remains to be seen. And time is short. Killing apes raises many questions. Is it murder to kill an animal whose DNA is so nearly identical to ours? Should cultural sensitivity be set aside when the risk of losing our closest kin is imminent? Ammann points out that no one seems to mind that in the big safari parks of East and South Africa, military-style guards patrol from airplanes and Land Rovers and don't hesitate to shoot poachers with high-powered rifles. Why, then, do we find this practice repugnant for ape poachers in the forests of western and central Africa? The conundrum leaves some in the conservation community scratching their heads.

BUSHMEAT AND DISEASE RISK

You can provide your children with much-needed protein by hunting and feeding them bushmeat. You can also kill them with such kindness. Meat as a food source is undeniably rich; a dense package of protein, saturated fat, and amino acids. But it comes with a host of pathogens; bacteria, protozoans, worms, viruses, and prions. We all know about the parasites. There are cafes in which patrons are asked to sign a waiver of liability before ordering rare steak or burgers due to the risk of *E. coli* infection. In theory the risk of parasitic infection from bushmeat is correlated with the degree of hygiene used in the butchery.

But since that level is likely to be low to nonexistent in a rural setting in Africa, one can presume that bushmeat is routinely consumed with an unhealthy dose of bacteria. If you roast or smoke the meat thoroughly but then chop it up with a knife used for butchering all sorts of game, you're likely to swallow plenty of infectious organisms.

Although the most common infections are bacteria and parasitic worms, which are likely to cause transient or chronic but rarely fatal health problems, there are some lethal pathogens lurking in bushmeat too. Great apes and humans are made of so nearly the same genetic fabric that we share and transmit many diseases. Ebola infections may spread via the consumption of monkey carcasses, as has been reported for chimpanzees eating colobus monkeys. As we shall see in the next chapter, Simian Immunodeficiency Virus (SIV) bears a not-coincidental similarity in genetic structure to its human kin and likely descendant, Human Immunodeficiency Virus (HIV). This is likely due to its emergence in our species as a result of cross-species contact with the blood or bodily tissues of wild chimpanzees sometime in the past hundred years. Other equally lethal viruses such as Marburg hemorrhagic fever, which carries an eight percent mortality rate, can cause fatal internal hemorrhaging and, like Ebola, can spread to humans who come into close contact with infected victims.

Anyone who handles bushmeat can be infected by a killer virus. A recent study of SIV occurrence in western Africa showed that at least one in five monkeys for sale in food markets or as pets tested positive for the virus. Using the same machete stained with infected monkey blood or perhaps getting too close to a pet can result in transmission of the disease.

In addition to viruses and bacteria, there are prions. We all know about prions these days, thanks to Creutzfeldt-Jacob disease and its close kin, so-called mad cow disease (bovine spongiform encephalopathy). Although the former is a rare human disease most often acquired by inheritance, mad cow disease appears to have the ability to make the jump into a human host via the consumption of beef, or more specifically the brain tissue and marrow of cattle. Those who eat infected beef are also consuming prions, which are such primitive organisms, composed almost entirely of proteins (*prion* is a portmanteau of the words *protein* and *infection*), that they very nearly can't be considered to be living things. Prions were first discovered in another mad cow–like disease, kuru, which is found among tribal people of Papua New Guinea and is passed among them through the ritual eating of the brains of deceased family members. The barely alive status of prions makes them sinister, since they are also almost impossible to destroy through cooking. And the neurodegenerative diseases they spread are virtually always fatal.

SOLUTIONS TO THE BUSHMEAT TRADE

There are no real solutions to the bushmeat crisis. But there are ways to mitigate its devastating effects. These tend to focus either on somehow constraining the supply of bushmeat that comes out of the forest, or on attempts to lessen the demand and the pressure on wild ape populations. Bushmeat, like all other commodities, exists in a supply and demand market. Conservationists therefore can try to mitigate the impact of the bushmeat trade either by limiting the supply or by lessening the demand, or both. Each is a fairly Herculean task. Placing limits on the supply of many commodities drives up their value, creating a black market even more profitable than when the supply is abundant. The most often-proposed plan to curtail supply is to enforce existing poaching laws with more and better guards posted along logging roads and in markets. The hunters who kill the game must sell it to local middlemen, who at least temporarily hold quantities of bushmeat before they can deliver it to market. But intercepting the meat in the hands of the middlemen before it makes its way to market is, although effective, labor intensive. Enforcement policies that currently exist on paper, if they exist at all, would have to be given teeth, and boots on the ground. The prospects for enforcement seem poor.

Wilkie and Carpenter propose that guards could set up roadblocks at which bushmeat would be subject to a stiff tax. This would require far less manpower, and in theory would drive down the middleman's profits while driving up the consumer cost in regions where cost increases can hardly be supportable by the local population. If the taxes are steep enough, perhaps even the cultural preference for bushmeat could be eroded. The intransigent problem that probably makes taxation unworkable, however, is the system of corruption that is so firmly entrenched in western and central Africa. Wilkie and Carpenter suggest that if the logging companies themselves were to take control of bushmeat taxes on their own land, they would succeed in limiting the trade because the companies maintain a corps of loyal guards. These guards would be less receptive to bribes than the local government-hired guards who would otherwise staff the taxation points.[12]

Another possibility would be to organize the system of bushmeat harvesting so that it is community-based and run by a local committee. A local committee could control collection and sales, allowing for adequate distribution without overharvesting within their jurisdiction. The fundamental problem with this noble idea is that African governments, not to mention businessmen, are unlikely to allow such a system to exist, let alone flourish.

By the same token, the current pressure on wild animals could be lessened by mitigating the need to harvest wild populations. Game ranching is already a big industry in some parts of Africa for the raising and sale of live animals for captive breeding, and the hunting of them as trophies. Such ranches also sell meat from their stock, but this is rarely their primary income because it does not generate enough profit. But the general principle of producing substitutes to bushmeat and offering them to hunting populations at prices lower than bush-meat commands is sound. An obvious example is domestic live-stock, but there are a wide range of animals whose meat might supplant that of wild animals, including great apes, in rural Af-rica, without some of the limitations and problems of livestock production. Domestic animals are more often viewed by villag-ers in Africa as repositories of wealth and insurance policies for rainy days rather than reliable sources of protein. This may be because Africa has a long history of endemic diseases—trypano-somiasis from tsetse flies for one—that severely limit cattle rearing.

The lowly rodent is a potentially viable replacement for at least some portion of the bushmeat currently traded and sold in Africa. Cane rats are fat and tasty animals that some Africans already include in their diet, and there have been attempts to breed them in captivity. Farm rearing has also been proposed for their more svelte cousin, the giant African forest rat. Rab-bits and porcupines can also be farmed, and there have been some nascent attempts to replace the bushmeat need with pork from either domestic or hybrid wild/domestic pigs.

Domestic livestock becomes a more viable substitute mainly when the supply of bushmeat falls to such low levels that little choice is left. And by that time the goal of saving wild popula-tions may be lost forever. Unless economic incentives are of-fered, nothing will change the culture and the cultural taste for bushmeat. Successful village outreach programs and environ-mental education provide some hope, especially since the meat of great apes is only a fraction of the total quantity and range of bushmeat consumed in most areas. Persuading people to give up ape meat may be far more easily achieved than persuading them to give up duiker antelopes or monkeys, as gorilla and chimpanzee meat is consumed mainly as a delicacy and not as a staple. The disease prevention factor discussed earlier may also appeal, especially to educated urban Africans and to immigrants living in western countries with access to black-market meats.

Unfortunately, it's not possible to simply tell local people whose cultural heritage involves eating apes to stop doing so. The experience of legions of foreign aid workers is illustrative. Young, idealistic people invent, for example, a solar cooker that could replace the campfires in use in villages all over the developing world, thereby saving precious fuel wood and lessening the need for women to spend their waking hours walking in search of fuel. They bring the solar cooker to a village and spend a day showing local people its wonders. And then they move on to the next village, whereupon the village women throw the cooker into a corner of their house and return to the time-honored practice of burning wood or cow dung or whatever else is available for fuel. If the aid workers speak the local language as natives, the odds of making a persuasive case for conversion is greater, but still, a new cultural tradition imposed from the outside rarely succeeds, even when the economic incentive exists. People, like all organisms, are conservative when it comes to innovation.

As with other conservation solutions, the most reasonable and best hope is not to stop all consumption of bushmeat but to mitigate it to levels at which great apes are not at extinction risk from the practice. Otherwise, even the forests that are protected from logging will be empty by the end of this century.

Chimpanzees, along with the closely related bonobos, are our nearest living kin and the most abundant of the living four great apes. *Photo by author.*

The global population of mountain gorillas, the largest primates on Earth, is around 750 and climbing. *Photo by author.*

The bonobo has been much hyped as the "make-love-not-war" ape. Its numbers are believed to be in the low thousands, all of them in the war-torn, troubled Democratic Republic of the Congo. *Photo by Don Byrd.*

Orangutans in Indonesia are in steep population decline, especially on the island of Sumatra, where only a few thousand survive. *Photo by Eric Fell.*

FOUR

Outbreak

The gorillas had been walking all day, stopping occasionally to climb a tree to search for fruit or plunder a saladlike growth of favored plant leaves. They were a group of a dozen, with one massive silverback male, another adolescent "blackback" male, and several females and their babies. During rest times, mothers watched their infants, young males jostled one another during play bouts, and a silverback kept a watch for danger, especially other silverbacks. After a long midday nap, the group walked south toward a stand of trees it knew well. But along the way, a female was diverted by an odor in the undergrowth. She investigated cautiously as the rest of the group brought up the rear. There, lying under a bush, was the carcass of a gorilla. The entire group grew agitated and made nervous calls as the silverback came up and began to gently poke the dead gorilla. They had all seen death many times, but it was no more comprehensible to any of them now than it ever was. The dead gorilla was, unknown to the silverback, one of his rivals over many years, and he had been dead for just a few hours. As the group peered intently at the body and glanced nervously around them, they trod on the leaf litter around the carcass. After an hour of nervous examination, they lost interest and walked on in search of that favorite fruit tree.

Within a few days, two of the group's infants and one of the females sickened and quickly died. Others became weak and showed signs of some sort of severe illness. Within ten days of their unwitting encounter with death, all the members of the gorilla group were themselves dead.

This scenario has played out many times during the past several years. In some forests in central and western Africa, nearly all of the gorillas have disappeared. The only clue to their annihilation has been carcasses littering large areas of rain forest. Chimpanzees have suffered the same fate in the same forests. The gorilla researcher Magdalena Bermejo and her colleagues, working in the Lossi Sanctuary in northwestern Republic of the Congo, began finding dead apes in 2002. They estimate that up to five thousand gorillas died in their study area during 2002 and 2003.[1] That represents a mortality rate of between 90 and 95 percent of the local population. It also represents the death of as much as 5 percent of the entire species in Africa. Among the 143 gorillas known as individuals to the researchers, 130 disappeared and were presumed dead. Surveys carried out the following year showed clearly that within a well-defined perimeter, the huge gorilla population of the region was essentially gone. Perhaps most tragically, the gorillas affected were among those least impacted by the other threats the species faces, such as logging and poaching.

Tissue samples taken from the bodies revealed the killer: a strain of the Ebola virus known in human outbreaks as ZEBOV. A year before the gorilla die-off, there had been an outbreak of the same Ebola strain not far to the west along the Gabon-Congo border.[2] Ebola is an eye-catching threat. Hollywood and the media have hyped the disease, but in fact, viral outbreaks like Ebola are like wildfires. The virus breaks out—from sources that are still unclear—then kills a majority of victims very quickly via massive hemorrhaging and organ failure before disappearing, usually before health care workers and researchers can arrive on the scene. It's a largely self-contained epidemic, the lethality of which is offset by its limited potential for widespread infection in humans. The maximum human death toll from any single outbreak of any Ebola strain has been in the hundreds. So the deaths of thousands of gorillas, not to mention likely thousands more chimpanzees, are a watershed event in the epidemiological history of Ebola as well as in ape conservation.

Outbreaks such as this one leave more questions than answers. What is the source of the virus? Did it spread because of some massive infection from whatever animal host was the source, or did the gorilla victims spread the disease from one to another? The natural source of Ebola was not pinpointed until recently. Early studies suggested that colobus monkeys might be carriers—the chimpanzee appetite for monkey meat would have explained in that case how Ebola has afflicted some chimpanzee populations. The current thinking, however, is that fruit bats are the natural reservoir, three species of which may exhibit the virus without suffering any harm from it themselves. A chimpanzee foraging in a tree might capture and eat a sleeping bat, but how the virus makes the leap from bats to vegetarian gorillas is not altogether clear. The symptoms of the epidemic in Lossi, however, suggested that it was not simply the result of massive infection from bats. According to Bermejo and colleagues, the deaths spread sequentially from north to south within Lossi. The cycle of deaths followed a well-established pattern seen in human outbreaks: a time lag of about eleven days between infections noted in gorilla groups living near each other was almost identical to the spread of Ebola seen numerous times in African villages.

It seems clear that at least in this one case, a lethal virus spread through the gorillas of northwestern Congo as a result of contact among gorilla groups. It's easy to see how the virus could spread within groups, from mother to infant and during sex. Gorillas are not territorial, and groups encounter one another in the forest on a regular basis. Even gorilla and chimpanzee groups have the potential for cross-infection. Gorillas and chimpanzees are known to feed in the same trees and encounter each other in the forest on occasion. There is every reason to think that a gorilla would show as much interest in a chimpanzee carcass, or vice versa, as either would display to a dead member of his or her own species.

As sensational as an Ebola outbreak may seem, it appears to be only the tip of an infectious disease dagger poised over many great ape populations. Great apes have enough threats stacked against them without biological weapons added to the mix. But that is exactly what many populations face. Much evidence indicates that diseases, some with human origins, are playing a significant role in ravaging great ape populations in Africa. The disease risks are not necessarily anthropogenic. Ebola may be transmitted from a naturally occurring small mammal, and other ape diseases have natural cycles that periodically subject ape populations to outbreaks, from which most subsequently recover. But because great apes and humans are so similar genetically, our infectious diseases are their infectious diseases. Influenza, pneumonia, tuberculosis, and the common cold can all be easily spread from people to chimpanzees. As indigenous peoples living deep in the rain forest discovered, first contact with missionaries and their diseases was a health disaster. Illnesses that western populations had lived with for centuries killed entire villages of local people with no acquired immunity to them. The same is true of apes who encounter human diseases.

Some researchers believe that the Native American population of the modern United States was more than fifty million in 1492. One hundred fifty years later, eighty percent had died of diseases introduced unwittingly by the European colonists. Europeans brought their own diseases, and the African slaves they brought to the New World also carried their own complement of diseases. Together, these created an onslaught of contagion that depopulated the New World of its earlier inhabitants as effectively as any biological weapon of mass destruction.[3] Indeed, some scholars argue that settlers knowingly infected native peoples by trading blankets and other items known to come from victims of killer diseases. Native Americans died of smallpox more often than any other epidemic, but measles, flu, bubonic plague (which itself had wiped out much of Europe's population in earlier centuries), cholera, mumps, and whooping cough took a terrible toll. Most of these diseases were less virulent than Ebola. They tended to sicken but not immediately kill the victim. The effect of this onset pattern was to allow one infected person to infect others before succumbing. The overall effect was rapid-onset illness and death.

The analogy of missionaries and settlers infecting native people accurately reflects how people infect wild apes. Not all ape diseases come from humans, but as humans encroach on ape habitat, infection has become an increasingly ominous threat. Great ape populations have been hit with epidemics since the first field biologists were able to monitor the health of wild animals. In some of the outbreaks, human sources are strongly suspected. In the half-century of research on Jane Goodall's famous Gombe chimpanzees, at least six disease epidemics have occurred, each taking a terrible toll on the community. In 1966, polio broke out, a crippling viral infection that was once a scourge of all humanity and until a few decades ago still a health threat in the developing world.[4] The epidemic may have spread from the human population in the villages surrounding the chimpanzees' home in Gombe. Goodall first witnessed the outbreak when an infant, Grosvenor, suddenly lost the use of his limbs and quickly succumbed. Polio vaccine was flown into the sanctuary and administered to the wild chimpanzees in bananas. But the deaths continued, including two of Goodall's most beloved chimpanzees. Six chimpanzees—about one in seven members of the whole population—died, and others were left with lifelong disabilities. Victims who survived the initial outbreak were left unable to keep up with the community as it foraged for food and were also victims of violence from other members of the community, who took advantage of the sudden helplessness of the afflicted.

Just two years after this tragedy, the Gombe chimpanzees were hit by a respiratory infection resembling pneumonia. Four more chimpanzees died, including David Greybeard, the old male whose approachability had allowed Goodall to penetrate the inner world of the Gombe chimpanzees. Respiratory flulike outbreaks occurred four more times from the 1980s to the early 2000s, each time causing the deaths of chimpanzees. Those who work to protect and manage the chimpanzee population at Gombe consider disease to be a major risk to the long-term survival of the community.[5]

Why have Gombe chimpanzees been the victims of epidemics so frequently? One reason may be that at Gombe, with its high-profile status and large number of researchers over the years, the health status of the chimpanzees may have been reported in more detail. In other words, perhaps Gombe is no different from any other chimpanzee population. Taï National Park in Ivory Coast has, for instance, been hit by a series of respiratory epidemics resembling commonly seen human respiratory infections, plus an outbreak of Ebola. But there is another possibility: Gombe is a tiny island of forest amid an ocean of farms and villages in western Tanzania. Contact with people, livestock, and the detritus of civilization may be less avoidable than in other, more remote chimpanzee study locations. Gombe's western edge is the vast shore of Lake Tanganyika. But to the immediate north lies a village, Mwamgongo, from which people and livestock occasionally wander into the forest. To the south and east lies an endless succession of scattered villages.

Until the late 1990s, fishermen were allowed to camp on the beach at Gombe for a portion of each month to ply their trade, and the ensuing camps and their trash contributed to the risk of disease transmission. Even when the fishermen were not present, chimpanzees would leave the forest to walk into the abandoned camps to search through the rubbish, lick the ashes from fire pits, and in general bring themselves into direct contact with the diseases that people harbor. The Tanzanian research staff also live in the park, although their families are no longer allowed to live there because of the risks that accompany human population density adjacent to a small forested area. In former times the staff and their wives and children lived in the camp, and public footpaths bisected the forest, allowing local people to traverse the park from the lakeshore to their farms and villages to the east. These public footpaths brought people into unintended contact with chimpanzees. Women carrying small children attracted the curiosity of the Gombe chimpanzees in particular, perhaps because human babies are viewed by wild chimpanzees as potential meals, just as they would view monkeys, piglets, and other helpless mammals.

I was once standing in the forest, watching chimpanzees with one of our research assistants, when a woman friend of his approached. She was walking on the public footpath but left it when she saw us, and strolled through the forest just a few meters from the chimpanzees. As we spoke, I heard a cry from inside a bundle of cloth wrapped around her back. She was carrying an infant in the African way, wrapped inside a *kikoy*. The second the baby's cries were audible, a female chimpanzee, Gremlin, descended her tree and rapidly approached the human mother. The field assistants and I interposed our bodies in between Gremlin and the lady, and with some difficulty managed to keep them apart until the woman fled the scene. Whether Gremlin intended harm or just was intensely curious is hard to say, but it was a situation that no doubt repeats itself often wherever humans and wild apes are brought into contact.

The researchers themselves are few in number and are required to follow guidelines about maintaining a distance from the chimpanzees at all times to lessen the risk of contagion. Even so, they pose a health risk to wild apes and their habitat. We enter the apes' habitat, and in doing so we change their lives forever. We alter their natural behavior to some small extent. The Gombe chimpanzees changed their travel patterns in the 1960s to focus on the bananas that were put out to persuade them to stay around the clearing where most of the observation and filming was being done. The banana provisioning was largely stopped a few years later, but the travel patterns were changed forever. This was a small interference, and in the big picture all for the good. The pioneering research that resulted led to a new perspective of the chimpanzees and also a new conservation ethic toward them. But it altered their behavior nonetheless.

Today, chimpanzee researchers at most sites have adopted new guidelines to mitigate the risk of infecting the animals we are trying to study and protect. An important study by Sophia Kondgen and colleagues points to researchers as the vectors for respiratory diseases affecting at least one study, the Taï chimpanzees of Ivory Coast.[6] Researchers are quarantined from the animals for at least several days upon arrival at a research station, so that human-borne infections might either reveal or resolve themselves before any contact occurs with the study subjects. Respiratory infection transmission is the main concern. When approaching chimpanzees in the forest a distance of thirty feet is maintained between apes and researchers. We don't eat, drink, or relieve ourselves in the forest vicinity of the chimpanzees, for fear that any of what we leave behind may infect a curious ape. Recently researchers have even begun to don surgical masks when in proximity to chimpanzees to lessen the transmission of airborne infections. With respect to our subjects, we are not unlike those missionaries in former times who risked spreading western diseases to defenseless indigenous peoples. They allow us into their lives and are then at our mercy.

Tourists also transmit diseases. People arrive in droves at sites where the prime attraction is observing wild chimpanzees, gorillas, or orangutans. It's expensive and logistically challenging to get there, and for many it is a lifelong dream fulfilled. Some ecotourists arrive worried about their safety in the face of a charging silverback gorilla or a displaying alpha male chimpanzee. The real safety risks are borne by the apes, not the humans. Tourists bring their own diseases with them, and sometimes they are passed on to the animals directly, by sneezing, or indirectly, through the leavings of tourists in camps that the gorillas or chimpanzees later explore. The established mountain gorilla ecotourism sites all ask tourists to drop out of their gorilla trek if they have a cold or the flu. Whether many do is unclear. We'll examine the risks and benefits of ecotourism in detail in Chapter 6.

Gombe epidemics have taken a severe toll each time, yet the population remains more or less the same, attesting to the ability of the animals to rebound from local outbreaks. Despite a series of epidemics at Gombe, the size of the main study community, the Kasakela chimpanzees, has not varied much over the past many decades. There has been little growth in its size, but also no population crashes or even negative growth. The trend at Gombe is unfortunately not matched by that of other chimpanzee sites, such as Taï in western Africa. There, an entire community was nearly wiped out by a combination of disease outbreaks and killings by local people. Anne Pusey and her colleagues examined the trends in community size at Gombe and argued that the chimpanzee population there has maintained its numbers despite disease outbreaks because, unlike the other sites, Goodall has often medicated chimpanzees who turn up ill in her research site. They also suggest that an increasing availability of good food and an influx of immigrants over the past few decades have served to inflate chimpanzee numbers in the face of epidemics.[7]

Ebola is the most sensational of the disease outbreaks to afflict wild apes and may have the most widespread devastation potential, but outbreaks of this virus have been less often recorded than many other garden-variety diseases. Over the past five decades, great ape researchers and health monitors have noted outbreaks of polio, influenza, a variety of respiratory infections some of which resemble viral or bacterial pneumonia, sarcoptic mange (also called scabies; a skin disease perhaps carried by mites and perhaps transmitted from local domestic animals), insect-borne diseases such as malaria, dengue fever, yellow fever, and a host of poorly understood viruses. In addition, we have seen an AIDS-like disease that appears to be caused by the ape counterpart and progenitor of HIV, and anthrax. This panoply of ailments are not necessarily all transmitted from humans to wild apes, but many clearly are.

Some of the diseases are naturally occurring in wild populations, and there is no doubt that left to their own devices, unhabituated, uncontacted wild chimpanzees would still die periodically in epidemics. During the rainy season, many chimpanzees in a given community show classic signs of colds and other viruses—runny noses and lethargy—that are no doubt the result of lowered immune systems during inclement, chilly weather. But many of the diseases listed above stem directly from contact between wild apes and humans, livestock, and other domesticated animals that venture into the forest, or the debris that humans deposit in or near the habitat of wild apes. And as with the Ebola outbreaks described above, it takes only one wandering ape who contracts a human-borne disease at the edge of the forest to pass the disease on to members of his community and beyond deep inside the forest.

At least two anthrax outbreaks have been recorded among chimpanzees, one of which also affected gorillas. Outbreaks in 2004 and 2005 in Cameroon were further evidence of the disease in wild great apes; the Taï chimpanzees had been hit by the disease earlier. Its status as a potential bioterror weapon made the outbreak seem terrifying, but in fact anthrax is a widespread livestock bacterial disease, often spread through water or food. Its spores can spend years lying dormant, waiting to be passed to a new host.

ARE FIELD RESEARCHERS KILLING THEIR SUBJECTS?

Given what we have now seen about the prevalence of disease outbreaks among ape populations that are well studied and have an ongoing presence of researchers, it's reasonable to ask whether our presence in the forest with the animals is doing more harm than good. It seems clear that all chimpanzee populations that have been studied closely by people have experienced periodic disease outbreaks. Indeed, respiratory disease is the main illness and mortality risk found among chimpanzees who have been habituated to human presence, whether for research or tourism. This is well documented at the three longest running chimpanzee study sites: Gombe, Mahale, and Taï. In at least one of these sites, Taï, epidemiological and genetic data show that respiratory infections have been transmitted to the apes by researchers. This suggests that having people approach apes to within a close distance, which is so central a part of great-ape field research, may be dangerous to the apes. This creates a major dilemma for scientists and conservationists alike, because the up-close-and-personal nature of human and habituated ape contact is essential to ecotourism too. Tourists pay large sums of money to see wild apes at close range, and that revenue is an important factor encouraging local African governments and people to preserve rather than destroy the forest and its apes. Moreover, the presence of research and tourism projects is a primary barrier to poachers, who would otherwise be preying on some of these ape populations. In other words, if you can't get close enough to the apes to see the details of their lives out of concern for disease transmission, the whole structure and operation of ape conservation may be at risk.

We must believe that field research on great apes is on balance a boon for the apes rather than a bane. The information gleaned from their lives and disseminated to conservationists is what allows policies and strategies of protection to be devised. The research, when described to an eager public, is also the basis for everything from conservation fundraising to ecotourism. Without the impetus of the research that normally precedes conservation, there would in all likelihood be no effective conservation. However, the law of unintended consequences is at work, and we are just beginning to understand the full effects of our own diseases when passed on to the apes. For instance, Peter Walsh, a conservation biologist, has argued that birth cycles in habituated chimpanzee populations have been skewed because of die-offs of infants from respiratory outbreaks that are periodically spread from human to ape, killing whole cohorts of the next generation.[8] Great ape researchers do not like to think that they may be influencing the demography of the entire population under study, but in some cases it appears that they have done exactly this. Nowadays, most great ape projects have their own veterinary aspect, and sometimes even a veterinarian on staff or on call. Veterinarians tend to be very hands-on, and field biologists and conservationists tend to be very hands-off, which creates a mild cultural conflict when decisions must be made about an animal's health. Should we tranquilize apes for the purpose of medicating them, or expect that with protection but not intervention, they have the means to heal themselves? The two sides generally respect one another's perspectives, and both understand the urgent need for disease prevention and control in great ape study sites.

AIDS AND GREAT APES

If you want to understand the origins and nature of a disease, it pays to think like the pathogen. This is central to the emerging field of evolutionary medicine, which argues that understanding a disease's reproductive and life history strategies will likely inform conventional medical researchers about how the disease may be more effectively treated. In other words, medical researchers ought to spend more time thinking about why a disease exists and why and how it evolved, in addition to worrying about how to treat it. This is much harder to do than it might seem.

For example, we expect that pathogens have evolved strategies that are very specific to their intended hosts. Their survival depends on living in lockstep with the biology of the host species that they infect for some portion of their life cycle. There is a relationship between the ease of transmissibility and the maximum virulence. The common cold is very easily passed from one human host to the next, but it's not going to kill a healthy person or even render you immobile for long. At the other end of the spectrum, HIV (human immunodeficiency virus) will eventually kill its host by inducing a range of immune system malfunctions. However, unlike the common cold, HIV is not easily transmitted. It requires direct fluid transfer, most often via sex. And since not every sexual encounter will result in transmission of the virus, HIV's transmissibility is actually quite limited, in inverse proportion to its virulence.

We now know that the lentivirus known as HIV entered the United States in about the late 1970s, but the virus had emerged in the human species somewhere around a century ago, give or take twenty years. Lentiviruses make a living by attacking the immune system of the host with a strategy of eventually debilitating it, but hopefully (from the virus's perspective) not before it can be passed on to a new host. This lentivirus later mutated into a variety of strains, but in its earliest inception it was a different viral strain altogether, SIV, or simian immunodeficiency virus. There are multiple strains of SIV, just as there are of HIV, but certain strains of SIV so closely resemble the two main types of HIV—HIV-1 and HIV-2—that it's clear SIV is a genetic relative of HIV. The most virulent, dangerous strain of HIV is HIV-1, and it has a close counterpart in the strain of SIV—SIVcpz—found in chimpanzees. Researchers Beatrice Hahn and Paul Sharp first made the discovery that chimpanzees were infected with a chimpanzee version of the virus SIV, which had made the leap into the human species and mutated into HIV-1.[9] Where did the chimpanzees acquire the virus in the first place? There are dozens of varieties of lentiviruses found in African forest monkeys; the red-capped mangabey and the spot-nosed guenon, both tree-living monkeys of West African forests, are likely natural lentivirus reservoirs. Chimpanzees, who are avid hunters of monkeys and consumers of their meat, likely became infected by eating these monkey species, whereupon the virus strains hybridized and later crossed over into our own species.

Exactly how this crossing over occurred is still a matter of intense debate, but the likeliest scenarios have narrowed to just a couple. The likeliest way it happened was when a villager in central Africa went out hunting and brought back a chimpanzee. All the African apes are considered highly desired table fare in central and western Africa, and bushmeat has included the meat of apes for decades, if not centuries, as we have seen. The hunter may have killed a chimpanzee and then butchered the carcass to feed to his family, or he may have sold or bartered the meat but used the same knife to cut up other foods that his family consumed. Or perhaps he brought a baby chimpanzee home as a pet for his children, and the SIV-positive infant passed the disease on to the hunter's children through a cut in the skin. In any case the simian version of what was to become HIV/AIDS made the jump from the ape to the villagers. This sort of cross-species transfer scenario should not surprise us. When we hear of flu outbreaks from Asia labeled "swine flu" or "avian flu," it's a reference to other cross-species infections. Swine flu outbreaks may occur when infected pigs are in close contact with people, such as swine industry workers, or when people live in very close proximity to their pigs, as in Hong Kong and China. Pigs are also frequently hosts to strains of avian flu, multiplying the risk of transmission of some strain or another to humans in contact with them. The 2009 global epidemic of H1N1 swine flu, though milder in severity, was not unlike the great flu pandemic of 1918, which is also believed to have been swine-originated.

So the SIV virus may have originated in an African monkey, been acquired by chimpanzees, and then passed on in mutated form to humans. Other, less likely scenarios have relied on western medicine intervening in the lives of innocent African victims. One such theory, now largely discredited, alleged that the polio vaccination trials of the 1950s, which were administered widely in villages in east-central Africa, may have spread the virus via cultured tissues taken from chimpanzee kidneys used to produce the vaccine. No trace of infection has been found in archival samples of that vaccine, however, and we now know that the origin of HIV in Africa far predates the 1950s. Others have argued that African laborers, victims of colonial exploitation, may have been the first victims during Belgian colonial rule in the region, or even that HIV was introduced to Africa in a deliberate act to kill black Africans. Both these ideas fail, again because phylogenetic studies of HIV show that HIV-1 appeared many decades before western colonial powers would have had the technology to introduce it to the local population. HIV-2's roots are less clear, but recent research suggests it arose in western Africa more recently, around the mid-twentieth century, and in the same region and time period as independence conflicts that were raging in former Portuguese colonies like the Republic of Guinea-Bissau.[10] Because the first non-African cases of HIV-2 were seen among Portuguese war veterans of that war, it seems possible that they carried the disease home, having acquired it from unsterile needles or from sexual activities.

Whatever the origins of HIV, it is clear that SIV has been around much longer and is an ancestor to it. In the late 1990s, at the time of the discovery of SIV, researchers considered it a major breakthrough in understanding HIV/AIDS simply to have identified SIV-positive chimpanzees in the wild. SIV occurrence in the Gombe chimpanzees was first documented in 1999. Occurrence was somewhat unexpected in pattern, appearing in various family lineages. No one seemed to think that SIV was more than a harmless virus in a great ape host; there was no evidence of deaths or illness or anything resembling AIDS symptoms. Then a paper published in 2009 changed that view. Researchers from the University of Alabama at Birmingham and elsewhere showed that among the same Gombe chimpanzees, those infected with SIV-cpz had a dramatically higher mortality rate. SIV-infected females had a lower birth rate and higher infant mortality rates. T-cell counts, the primary indicators in HIV victims of the body's efforts to fight immunosuppression, were severely depleted in SIV-positive Gombe chimpanzees.[11] At least one animal's necropsy showed symptoms closely resembling what one expects in a patient dying of late-stage AIDS.

SIV infection is clearly not harmless in wild chimpanzees but rather a potential killer with a mortality potential that may approach what we have seen in the AIDS pandemic. The distribution of SIV occurs in a patchwork pattern that suggests it does not have a deep history in chimpanzees and is not a naturally occurring virus for them as it appears to be in monkeys. Researchers speculate that since SIV appears to be less lethal than HIV, chimpanzees may have adapted to the virus to some extent. The realization that chimpanzees are under threat from SIV is fairly new to researchers, even though chimpanzees themselves have been suffering from its effects for some time. It is not yet clear whether gorillas, bonobos, or orangutans will be found to exhibit their own strains of SIV and what the consequences of that would be. But you can be sure that the next generation of researchers will be looking into exactly that possibility.

DISEASE AND THE ECOSYSTEM

Infectious diseases are taking a terrible toll on wild great apes. Some of these are naturally occurring, others occur naturally but can be exacerbated by human-aided transmission, and some are transmitted directly from human hosts. As humans change the landscape, we come into greater and greater contact with wildlife, including great apes, and pathogen transfer will only increase. The primatologist Colin Chapman and his colleagues argue that we should see the increasing risk of exposure to pathogens happening at three levels simultaneously. First, the local level; villagers clear forests to make farms and apes crop-raid. Tourism and research contact only add to the levels of local contact. Second, the regional level; people alter or destroy primate habitats, bringing them into contact with apes and increasing disease transmission. Third, the international level; global-scale human activities produce climate disruptions that play a key role in driving the emergence and spread of pathogens. Chapman and his team have shown that logging can lead to increased infection by intestinal parasites in forest monkeys and apes in Uganda.[12] In degraded forest fragments, the odds of contagion from human to nonhuman primate are far greater than in deep, pristine forest. The precise mechanism of infection is not always clear. It could be direct infection through contact with human settlements or indirect infection through lowered immune systems, induced by stress from poor nutrition or lowered genetic diversity.

The larger question is whether these infectious diseases can play a role in disrupting entire forest ecosystems, either indirectly through die-offs of primates or directly in some way we do not yet understand well. We've already seen the bitter irony that humans trying to study and protect apes can often expose them unwittingly to pathogens, causing illness and death. In addition to the direct, human-aided transmission of disease, our destruction of tropical forests in Africa and Asia brings wild apes into contact with human pathogens. Just as the first wave of European diseases hit Native Americans hardest, the first exposures of great apes to our diseases, before any immunities can be acquired, are likely to have the most devastating effects.

Transmission of disease from humans to apes occurs primarily at the boundaries between our two worlds: the forest edge. That is where the rubber meets the pavement in pathogen transmission. Gorillas wander from their forests to feed on farmers' crops, and perhaps rummage around in village trash heaps too. Villagers hike into the forest and shoot chimpanzees. Streams running between settlements and forests are bound to have cross-infection potential, and crop fields are irresistible draws to chimpanzees and gorillas and bring people and animals into conflict. Alien or domestic species, most often associated with people, introduce diseases against which apes have no immunity to their habitat.

Can infectious disease ever cause the extinction of an entire population or even species? Rabies has devastated the already-rare Ethiopian wolf, which has a total global population of under five hundred, to a point likely beyond recovery without dramatic human intervention. The same situation applies for island fox populations on the west coast of the United States, and to a lesser extent for African wild dogs. All wild canid populations are threatened by rabies spread from a variety of sources, but especially from domestic dogs. Canine distemper is in particular a killer of African wild dogs. In one of the most bizarre recent cases, Tasmanian devils were hit by an epidemic of a parasitic cancer that is infectious. Because genetic diversity in the immune systems of Tasmanian devils is very low, the disease can be passed from one to another through bites. Perhaps half the devils known to exist a decade ago have been killed by the disease, with no end in sight.[13]

One effect of pathogens is that they can mediate competition among species in a previously healthy ecosystem. When the North American gray squirrel was introduced to European forests, it began to outcompete the local red squirrels, much to the chagrin of Europeans. But when a disease, SQPV, hit the two squirrel species, the immunity of the gray versus the susceptibility of the red was a hammer blow to the red squirrel's ability to cope with competition from the American marauder. Similar interspecies differences in resistance to pathogens have resulted in tidal swings in ecosystems involving seals, canids, wild cats, and foxes. It's clear that the full effects of pathogens introduced to wild great ape populations go far beyond the apes alone. In fact, even the Ebola epidemics have the potential for causing mayhem in the future. Satellite data show that previous outbreaks of Ebola occurred during dry periods that immediately followed very rainy periods. If there are environmental events that can trigger outbreaks, perhaps by influencing the mortality or the behavior of bats and whatever other species may be carriers, then climate disruptions of the sort we are seeing ever more frequently in recent decades may bring about cyclical outbreaks of this and other hemorrhagic fevers.

The good news, if there is any, is that there may be critical lessons to be learned from research into infectious diseases affecting wild apes. When the genomes of two species are intertwined due to a similar history of dealing with a disease, studies of one can well inform our understanding of the other. Some of the infectious disease outbreaks mentioned above were used as models to better understand the SARS viral outbreak in humans a few years ago. Future generations of conservation biologists may be able to use genetic technologies they learn from SIV-cpz infection today to treat HIV victims tomorrow. Unfortunately, though, such benefits for our species, flowing in cases of diseases that we introduced to wildlife leading to their eventual extinction, may ring hollow to conservationists. What we need to know today is how to contain disease outbreaks that are wreaking terrible havoc, sometimes seen but most often unseen, on wild great apes everywhere.

FIVE

In a Not-So-Gilded Cage

The city of West Covina, California, a suburb of Los Angeles, had a mascot, and his name was Moe. Moe was an adorable child with the scrunched-up face of a wizened old man and the clumsily affectionate nature of a toddler. He was raised by St. James and LaDonna Davis, a loving couple who were his adoptive parents. St. James claimed he had rescued Moe from a poacher during a trip to Africa. In all likelihood, Moe had been taken from his mother after she was shot by poachers and ended up in the hands of an animal dealer who sold him to the unwitting Davis. Moe attended supermarket openings, civic functions, and fundraisers. Most of the neighbors felt fortunate to have such a cherished icon of the community living so near them. To the community, Moe was a child frozen in time, with no memory of the past and no thoughts about the future.

But Moe grew up. He became an adolescent and then an

adult male chimpanzee. By the time he was twenty, he was too destructive to keep in the Davises' home, so they built him a large cage in the backyard. Moe broke out of his cage and ravaged a police car. A few years later he bit off the finger of a visitor who had inserted her hand through the bars of his cage. At this point, his days as a city mascot far behind him, the city officials of West Covina declared Moe a menace to the community and removed him to a nearby wildlife rescue facility. The Davises sued, and a settlement was reached: Moe could no longer live his life in suburban Los Angeles. He would be moved to a wildlife sanctuary outside the city limits, in Bakersfield, where the Davises could visit with Moe as often as they liked. But the story did not end well. Moe was housed by himself in an enclosure adjacent to a cage holding two other male chimpanzees. On one of the Davises' visits, during which they had brought Moe a birthday cake, the two male chimps next door broke out of their cage and savagely attacked the Davises. St. James bore the brunt of the attack; he was mauled and mutilated horrifically before the sanctuary could dispatch the two attackers with a rifle. When a wild male chimpanzee attacks an enemy, he most often does so with a frontal assault, biting any vulnerable extremity; fingers, toes, facial features, genitals. In the Davis attack as well as in a more recent attack by a pet chimpanzee in Connecticut, the method of attack was similar. It is just what male chimpanzees do when they feel territorial. Mr. Davis survived the attack with terrible disabilities, as did the victim in Connecticut. The story didn't end happily for Moe, either. A few years later, he escaped his cage in Bakersfield, or was perhaps liberated under suspicious circumstances, and was never seen again.

Moe's story is perhaps an extreme one among the thousands of chimpanzees who have been taken from the wild to become human companions. But it shares one fundamental feature with nearly every other story of apes in captivity. It began and ended unhappily for all parties. He was separated from his mother as an infant and brought into a world in which he utterly did not belong and in which he could no longer be a chimpanzee. He served a family as their surrogate child but remained always a highly intelligent wild animal, with all the attendant emotional and psychological needs. These needs can never be met in a human household the way they are met in a social group of chimpanzees, any more than a person would survive psychologically intact if forced to live isolated from other people in a cage for twenty years.

Animals don't have rights in the legal sense any more than babies have such rights. To have a right to exist free from suffering and harm, you must be a fully sentient being capable of expressing that need. Both courts of law and philosophers of bioethics have tended to agree that nonhuman animals as well as very young people don't have the cognitive capabilities that warrant legal rights. Instead, society owes all creatures, human or nonhuman, a guarantee of safety from harm. Welfare. It's not just a semantic difference. We have a moral obligation to protect and defend the welfare of all creatures, one that supersedes the vagaries of the legal system of any one society. This moral obligation is too often overlooked, or more often rationalized. When a social worker removes a child from an abusive home, that's child welfare. And when we work to remove a chimpanzee from a six-by-six-foot laboratory cage or from a filthy zoo enclosure, that's ape welfare.

Why the fuss over definitions? Imagine you're a scientist, and you're seeking a cure for a disease for which you need experimental subjects very similar to humans on which to test your cure. So you propose injecting human subjects with experimental drugs, in what would be a gruesome biomedical experiment. You would use people taken from institutions, using only those whose IQ fell below a certain level, and who were incapable of speech. The immorality and illegality of this practice is obvious. We might think it would never happen today (although it certainly did happen often in an earlier era in American history, and it still happens in China and elsewhere). But if the same scientist wanted to conduct his study using chimpanzees, until recently there would have been no uproar. This is despite the inconvenient fact that a cognitively healthy adult chimpanzee could be more mentally able, however we choose to define intelligence, than a cognitively impaired human. The ape might even have communication skills the human lacked. So the societal choice to spare one species from torture while exploiting the other is fundamentally irrational.

In theory, our nearly identical genetic makeup should make a chimpanzee the perfect nonhuman experimental model. But decades of research have shown this not to be the case. For every human disease that can transmitted to a chimpanzee—hepatitis A and C, for example—there are others, like malaria, hepatitis B, Alzheimer's, and AIDS, that can't. Breast cancer is not a disease that afflicts female chimpanzees, and male chimpanzees rarely if ever develop prostate cancer. This leaves a large gray area in which the uncertainty of how well drug studies done on an ape would apply to people warrants erring on the side of humanity.

The overall number of great apes in captivity in the world is tiny compared to their numbers in the wild, although this may change in coming decades. So you may wonder why I'm bothering to include this chapter on the fate of captive apes in a book about conservation and extinction. It's because the ethical issues of maintaining great apes in captivity—whether for biomedical reasons, for basic behavioral research, or to give them sanctuary—have implications for how future generations care for both wild and captive populations. I believe we should be far more concerned with the welfare of natural ecosystems and the communities of wild animals that inhabit them than with the life of any one captive or wild animal. Life in the wild is not sacrosanct; it is given and taken away every millisecond. We never want to cause unnecessary suffering. But when the needs of one animal—be it a grasshopper or a gorilla—conflict with the exigencies of preserving the natural habitat in perpetuity, I believe we should always come down on the side of the environment. There's also a fundamental distinction between the moral obligation to protect naturally occurring wildlife and their habitats—which is all important—and protecting invasive nonnative species or domesticated animals. The welfare of the animal—no matter how adorable or easily anthropomorphized—cannot compare to the need to protect the natural habitat of wild animals whose existence is often compromised by human-introduced invaders.

Some animal rights groups protested the shooting of burros from the Grand Canyon, though they are human-introduced pests that wreak environmental havoc. So burros are removed today humanely, in a good compromise of the needs of the environment and the sentiments of people. The same groups decry the removal of the many goats, cats, and pigs that were introduced to the Galapagos Islands and which are now being eradicated by sharpshooters. They object to deer and duck hunting. I find such protests deeply misguided, both morally and environmentally. We attempt to right the wrongs that humans created by eradicating invasive species, by culling animal populations that have boomed so dramatically that they overrun and destroy native plants, and by allowing sustainable, regulated hunting of some species.

For every captive chimpanzee who seems well-adjusted and contented with its owners or trainers, there are many who have languished and either died or been consigned to locked cages because they were deemed dangerous. And of course these are just the captive apes lucky enough to be raised as human companions rather than living as pincushions for experimental drugs or as test subjects in the space program. There are currently somewhere between a thousand and fifteen hundred chimpanzees in labs and zoos in the United States. Some of these are experimental subjects in biomedical laboratories, others are former subjects now rescued to sanctuaries to live out their days unmolested. In no case is theirs a happy fate. There are increasingly few chimpanzees in captivity in North America who were taken from the wild, since that practice stopped several decades ago. The vast majority were born to parents in a cage or zoo enclosure and know no other life than the stupefying boredom of imprisonment.

Great apes are raised in a captive environment rather than growing up in the wild for several reasons. They may be incarcerated (the term is not inappropriate here) for the purpose of biomedical experimentation. They may, like Washoe, be in captivity for noninvasive studies, such as those done by psychologists interested in language. They may be in zoos or kept as household pets. These days, they are virtually all the offspring of other captive chimpanzees rather than imports from Africa. Such imports were halted more than thirty years ago. From a conservation standpoint, they are genetically dead once they or their genetic lineage leaves an African forest. One can argue—albeit weakly—that great apes in zoos serve an educational purpose in teaching the public about the extinction crisis facing them. But even when they are serving this cause, keeping apes in cages and confined is a thorny ethical issue.

APES IN BIOMEDICAL RESEARCH

It's hard for those of us who have spent years watching wild great apes to imagine what goes on in the minds and hearts of those who conduct medical experiments on them, even when such studies are intended to eventually help human patients. I suspect this is because as scientists, we know that any field of research, whatever its merits, is an industry. For every study that produces breakthrough results or important new theories, there are hundreds that amount to little more than background noise. They've been deemed worthy of funding after a review by peers who themselves are terrified of losing their own research funding. They will help some faculty member get tenure or a salary increase. Unfortunately in the case of experimentation done on great apes, many sentient beings are tortured or die in the process.

In 2006, New Zealand became the first nation to legislate legal protection for great apes, prohibiting the use of all great ape species in laboratory experimental research.[1] Their use in teaching was also banned, except in exceptional circumstances. This was in view of their psychological and emotional status as too similar to humans to be subjected to the sort of lab treatment from which humans are protected the world over. Two years later, the Spanish and British governments weighed enacting similar legislation. And in 2009, the United States, a major consumer of chimpanzees in biomedical research for decades, considered the same protections.[2] The Great Ape Protection Act, congressional bill H.R. 1326, would ban the use of chimpanzees and other apes in invasive research and would end federal funding for research using chimpanzees. While the motivation of those sponsoring the bill is animal welfare, the rationale that may get the bill through Congress is a more practical concern; the astronomical cost of keeping chimpanzees in federally funded laboratories. The bill argues that ending invasive research on lab chimps could save nearly two hundred million taxpayer dollars by sending experimental animals into a life of retirement in animal sanctuaries.

What happens to apes in biomedical laboratory studies? I meet many people who think that infecting chimpanzees with human diseases is a thing of the past. That would be a comforting thought, but it is only partially true. Without years of outcries from those concerned about ape welfare, it would not be true at all. Some apes who appeared to have been rescued from labs may not have been rescued at all. Nearly two hundred chimpanzees owned by the U.S. government and originally part of the space program were housed at Holloman Air Force Base in New Mexico and were used extensively from the 1950s to the 1990s in medical research that included being injected with viruses, vaccines, and pesticides.[3] These experiments were justified, researchers argued, because they involved research for which no other animal model was akin enough to humans. The chimpanzees eventually were used to form a breeding program to produce ape subjects for more experiments, and the animals were leased to a man named Frederick Coulston, a toxicologist who withstood enormous public pressure and began a whole new round of sticking the chimpanzees with insecticides and chemicals intended for use in women's cosmetics and perfumes. Through years of legal wrangling, Coulston became the owner of the two hundred hostage apes and their offspring. He moved them to a new facility where, despite new federal standards for ape cages, they were housed in cramped concrete cells. Death rates soared, and disease and infection rates were far above the norm for other captive chimpanzees. Coulston steadfastly fought animal welfare advocates who used the new Animal Welfare Act to try to bring some sanity to the situation.

Coulston eventually declared bankruptcy, and some of the animals, now numbering nearly five hundred, were moved to a "retirement" sanctuary in Florida where they could live out their days free from invasive experiments. But nearly two hundred remained at Holloman Air Force Base. These last experimental subjects, rather than being freed from further harm, are slated to be sent to the Southwest National Primate Research Center in San Antonio, Texas, where they face the prospect of more testing.[4]

This is all despite the fact that as research subjects for many of the most urgent and intransigent human diseases, chimpanzees fail miserably. Efforts to infect chimpanzees with the HIV virus in the 1980s in order to study the progression of AIDS in a nonhuman subject failed. Hardly any of the apes developed AIDS. As we saw in Chapter 4, they instead develop their own species-specific strain, SIV, or simian immunodeficiency virus. Moreover, the vaccines developed for HIV-1 infection, when tested in chimpanzees, did not either predict or correlate well with the expected disease onset in humans. Despite this, some scientists still call for a return to the use of chimpanzees in AIDS research. Malaria research using apes has also been deemed a failure, and even cancer research using chimpanzees is considered a boondoggle by many researchers. So why are chimpanzees in demand in biomedical research at all? It's all about the money, of course. The animal research industry in the United States is vast, and it's not going to surrender potential lab subjects or admit that they are fairly useless without a long hard fight.

Many advocates of ape welfare have fought to provide laboratory apes with more humane standards of care. The Animal Welfare Act, signed into law in 1966, was supposed to alleviate much of the suffering of lab animals in cramped cages. The act has been updated and amended numerous times since then, but its regulations for humane treatment ultimately fall back on the wording of the original legislative bill.[5] It is enforced primarily by the U.S. Department of Agriculture. While it is true that the days of housing a chimpanzee in a five-by-five-by-seven-foot cage are disappearing in the United States, they are not entirely gone. In many overseas laboratories it is still the norm. The wording of the Animal Welfare Act is slippery enough that enforcement is difficult and sporadic, and the interpretations of standards of humane treatment are vague enough to allow continued abuses. Great apes may be just the tip of the iceberg, being so few in number in laboratories compared to the millions of cats, dogs, and rabbits, but they also merit our deepest concern and the highest standards of treatment. The suffering and mortality of chimpanzees in the Coulston facility happened in spite of the Animal Welfare Act. Laboratory conditions that fall between the lines of what is spelled out in the act itself are left open to the interpretation of panels of researchers, who tend to take a less stringent view of preventing suffering of experimental subjects than the act intended. The USDA and other regulatory bodies are so chronically understaffed that offending facilities have no trouble avoiding enforcement. Laboratory workers who become whistle-blowers may receive a guarantee of legal protection if they are employed in government labs, but in the private sector no such protection exists, and whistle-blowing is rare.

Better treatment in labs has been a campaign waged largely by those outside the laboratory industry, even though it's those scientists and workers in the labs who have to face their victims every day. The Animal Welfare Act has specific wording only about the most basic aspects of housing and caring for primates, such as climate control, and leaves a great deal to the imaginations, and machinations, of lab directors and researchers.

My point is not to say that all use of experimental lab animals should end. It is to say that because of the special status that chimpanzees and the other great apes occupy with regard to mental capacity and emotional and psychological complexity, the use of such creatures in experiments should require un-equivocal, iron-clad evidence that they are absolutely indis-pensable, and that their use could directly lead to cures for human diseases. Extraordinary claims require extraordinary supporting evidence. No such strong evidence exists. Chim-panzees and other apes are socially complex enough that the concept of posttraumatic stress disorder applies to them. Do we want to use such animals as models on which to base our claims about normal human health?

APES IN NONINVASIVE CAPTIVE RESEARCH

Psychologists, biologists, linguists, cognitive scientists, and an-thropologists share a long history of considering apes to be the raw material on which careers are built. Of course there were always deeply intriguing and important research questions at stake, and there always will be. We no longer import wild-caught apes from Africa or Asia, all research subjects these days are born and raised unaware of any life other than their life in captivity. Some have argued that so long as they are humanely treated, why not explore the roots of human nature with benign studies of their behavior and cognition? After all, they're in captivity for life, and justifying the cost of their care may war-rant that some results, however indirect, be produced. Indeed, basic research results of just the past few years have provided astounding revelations. Many years ago it was shown that Ai, a chimpanzee in Japan who is part of the research program of Tetsuro Matsuzawa, had concepts of number sequences up to ten and could arrange the symbols (our written Arabic numerals) in proper order. Now another chimpanzee in the same program, Ayuma, consistently bests university students in short-term memory tasks, taking an instant mental snapshot of numbers flashed on a computer screen for a fifth of a second.[6] His recall of these numbers is twice that of the humans against whom he competed. Chimpanzees at the Yerkes Primate Research Center in Atlanta recognize kin who are unknown to them on the basis of facial similarities, just as we would.

Some of the most headline-grabbing research in science has involved captive apes. When psychologists raised a baby chimpanzee in a human household, providing her with all the enrichment that one of our own children would receive, the product was an ape who learned to understand and use language. When other psychologists have reared baby chimpanzees in laboratory settings, with trainers and students shuffling in and out with less of a semblance of being surrogate parents, they have ended up with apes whose linguistic skills are more limited. This led at least one researcher to claim that apes don't really learn language at all; they're just clever mimics. Ironically, he didn't consider the critical role of the learning environment itself. We know from recent research that interaction with human caregivers raises cognitive performance by chimpanzees in psychological research. It makes intuitive sense;

children raised in orphanages perform poorly on cognitive tests compared to those raised in loving homes with central care-givers.

The granddaddy of modern attempts to teach language to great apes is the Washoe study. In 1967, the psychologists Allen and Beatrix Gardner acquired a baby female chimpanzee and, with graduate students acting as surrogate parents, began to teach her language. Because chimpanzees lack the anatomical structures needed for speech—their palate and larynx are all wrong for clear enunciation of words—the Gardners chose American Sign Language as the medium in which Washoe would eventually communicate with her human family. After initially teaching Washoe language via molding her hands to form signs, the researchers realized the chimp was also learning on her own, picking up signs by watching humans model a behavior or an object by making the sign for it. This is the way that our children learn spoken language, and it led to simple conversations. Washoe was communicating by using learned signs and even inventing some of her own, not by mindlessly parroting signs she didn't comprehend.[7]

Washoe eventually learned to use more than two hundred signs and understood a far greater number. Until her death in 2007, she and her group mates lived for many years in a captive colony of language-using chimpanzees at Central Washington University, under the care and direction of one of the Gardners' former students, Roger Fouts, and his wife, Debbie.

There the Foutses demonstrated beyond a reasonable doubt that the chimpanzees spontaneously signed with one another, using the language they had learned from their human caregivers. Many other ape language studies followed, some with stunning results. A bonobo, Kanzi, became perhaps the most linguistically skilled ape ever studied. Koko the gorilla achieved fame with the claims, not all of them solidly demonstrable, of language comprehension and cognition.

A few of the studies failed to show clearly that the apes understood what they were doing. When Nim, a chimpanzee reared in a lab at Columbia University, failed to achieve the sort of linguistic competence that Washoe had attained (and which Kanzi and others would achieve later), Nim's researcher claimed he had exposed a fatal flaw in the ape language claims. Fouts and others pointed out that Nim was kept in a sterile laboratory, sometimes in a cage, and interacted with more than fifty lab assistants during his three-plus years as a research subject. As a result Nim likely had never had the opportunity to form the tight personal bonds that are so essential to healthy cognitive development in humans and in great apes.

My concern is not the debate over the results of the ape cognition studies. It's the fate of the apes in the studies. Throughout the early history of psychological research on apes, they were treated the same as all other animal subjects. They were disposable commodities to be discarded in the least expensive way upon termination of the researcher's funding.

This usually meant shipping apes and other primates to biomedical labs, where they would be subjected to gruesome experiments while confined in appallingly inhumane conditions. Even some of the apes used in the modern era of language studies ended up in such labs. Booie was a sign language–using chimpanzee who, after a career as a research subject, ended up in a biomedical laboratory. He was eventually rescued from this fate and retired to a sanctuary in California. Until that time his incarceration was little different than that of many other chimpanzees. Although he could articulate his thoughts and had committed no crime, he lived in isolation behind the bars of a cage as small as five by five feet, was subjected to biomedical experiments, and interacted only with humans in surgical masks.

Some researchers have accepted the moral obligation they feel to provide care for their research subjects rather than turn them over to medical experimentation. Roger and Debbie Fouts not only continue to conduct language research on the two chimpanzees still living from their long-term studies of decades earlier, they work to ensure that the animals, who may outlive the Foutses themselves, have lifelong care. This commitment of funds and time is obviously enormous and not something that many other scientists have been willing to consider. But it's the moral thing to do when your research career has ridden on the coattails of an ape's clever mind.

Daniel Povinelli is a researcher at Louisiana State University who has established a name for himself by studying chimpanzee cognitive development. Povinelli maintains that whatever chimpanzees think, their capabilities and thought processes parallel rather than reflect their close evolutionary kinship with humans.[8] Much of Povinelli's work has involved cross-fostering studies, in which baby chimpanzees are taken from their mothers and reared in human environments of the sort in which Washoe was raised, and which the Foutses and others now decry as wholly inappropriate. The vast majority of such cross-fostered chimpanzees exhibit classic symptoms of what in humans would be diagnosed as clinical depression and other neuroses, which presumably stem from the broken bond of mother and infant. Certainly human infants would be invalidated as research subjects if they were known to have been taken from their mothers at an early age and later exhibited signs of psychological stress.

Povinelli responded to widespread criticism of his research methods, and of Louisiana State University's lack of scrutiny of them, by shifting his cognitive studies to infants who were being reared in social groups with their biological mothers. The most resounding issue raised as a result of his work is not an assertion about human intelligence. The concern is that in the twenty-first century a researcher could make such assumptions about chimpanzee psychological complexity, or lack of it, that allowed the original cross-fostering studies to be undertaken in the first place. It's the modern-day equivalent of a blissful ignorance of the individualism of apes that Jane Goodall had to combat a half-century ago.

APES IN ZOOS

Jane Goodall is Distinguished Professor of Anthropology and Occupational Sciences and Therapy at my home university, the University of Southern California. The connection between Goodall's work and anthropology is obvious. As our closest living kin, apes are the subject of great interest by those scholars who specialize in understanding humans as a biological and cultural animal. But the connection to occupational therapy? Occupational therapists work with people in institutions, including prisons, whose psychological and physical health depends on their ability to enrich the endless hours of time they have on their hands. Apes in zoos are not so different from humans who reside in institutional facilities where there is no freedom and where they may spend every day of the rest of their lives.

I am not one of those people who claim zoos are evil. Many zoos are depressing, although it doesn't have to be that way. Zoos are perfectly appropriate and valuable facilities for the breeding of many endangered animal species. Zoos are also able to exhibit some of these animals to raise awareness and educate the public about global or regional biodiversity. Snakes and other reptiles, especially those hatched in captivity, have their physical and psychological needs met in many zoo enclosures. Many small animals are the same. But there are some animals that do not belong in zoos. Many birds—parrots for example—are as socially and psychologically needy as any ape, but we think nothing of sticking them in cages for decades, all alone. Many large mammals, either because their natural evolved behavior involves extensive travel or because they have psychological needs that surpass those of most creatures, simply do not belong in zoological parks. Elephants, big cats, wolves, and large-bodied primates like apes all have needs that cannot possibly be met well in captivity. They may live decades and reproduce well, but depriving them of their natural patterns of behavior through an impoverished social and physical environment is not morally acceptable from an animal welfare standpoint.

Ape welfare, and not ape rights, should be the bottom line. Animal rights advocates might argue that humans have no moral authority to place apes or other creatures in cages and that we cannot use apes to achieve a human desire, be it captive breeding, education, or entertainment. Animal welfare concerns are, by contrast, generally compatible with the concerns of conservationists. In order to achieve regular captive breeding the species' particular needs must be met, physically and especially socially. Zoos cannot function as global biodiversity arks; the amount of space available in the world's zoos combined is woefully inadequate for that purpose. But good zoos work with a limited number of species of great concern, and through organized networks of breeders, they cooperatively manage global collections of some of the world's most endangered animals. In the future, some of these species may exist only in captivity. Zoos can't save vital habitat, but they can serve a critical purpose of breeding for perpetuity, with the implicit hope that the conservationists are working to take care of the future. Such was the case when, over enormous objections and arguments, the last free-living California condors were captured and brought into captivity, the future of the species highly uncertain. Proponents of captive breeding argued that the species had no future in the wild. The number of the great birds had been reduced to only a handful. They admitted, however, that it was unknown whether the condors could be bred successfully in captivity. Meanwhile, the opponents of captive breeding argued vehemently that unless steps were taken to save the natural habitat in southern California, which was rapidly being degraded and outright disappearing, there was no reason to trap the remaining birds. But the let's-try-to-breed-them faction (mostly government wildlife biologists) won out over the let-them-die-with-dignity crew (the Sierra Club and other conservation organizations). Two decades later, the verdict is in, and the captive breeding proponents were right. Not only have hundreds of California condors been bred in captivity, more than a hundred are flying over their ancestral lands, including regions from which they had been extirpated hundreds of years ago.

This same principle won't work for great apes—they breed too slowly for any captive-breed-and-release plan to make sense. But zoos love baby apes. They're great for promoting the zoo, great at plucking at the public's heartstrings and giving out a sense of a conservation-minded zoo at work. Zoos had a long history of taking babies from their mothers, always "for the sake of the baby" or out of concern for "maternal neglect," but in many cases it was a thinly veiled effort to have a baby to put on display in the children's zoo or to parade onto local television shows in the name of marketing. That practice has been greatly curtailed in recent years, at least in the world's larger and better zoos. Captive-bred apes have replaced the need to import animals for zoo exhibits, but captive breeding offers no hope of turning the tables on ape extinction.

When it comes to zoos, size matters. The largest zoos in the United States, Europe, and elsewhere tend to be best in terms of animal treatment. They have the financial wherewithal to build better exhibits, provide better medical care, and house more individuals in large displays, offering the animals a richer social environment. Smaller facilities, especially in developing countries, are often run like something out of the nineteenth century, with chimps sitting in squalid, cramped concrete enclosures. On a recent visit to the Beijing Zoo, famed for its large giant panda collection, I walked up to a grassy outdoor island which, according to the plaque, housed chimpanzees. There were none outside, and clearly none had been outside for a long time, given the tall grass and overgrown greenery. The captives were inside, sitting in small, dark cellblocks running with urine and reeking of feces and food. A Chinese colleague later explained to me that the keepers prefer not to put the chimpanzees outside because maintenance and animal management are so much easier when they stay indoors. Occasionally, small zoos and even private individuals provide unparalleled care, especially when they specialize on just a few species and can devote themselves to care and breeding in a way that a huge overfull and understaffed institution cannot. But for the great apes and other large animals, smaller zoos should not be in the business of making money by keeping creatures whose health and happiness they cannot ensure.

APES IN ORPHANAGES AND SANCTUARIES

The men have been paddling their little canoe for hours, and as dawn breaks they reach the sandy eastern shore of Lake Tanganyika. They unload their cargo, which includes a couple of filthy sacks in which small bodies squirm. Sacks slung over their shoulders, they hike from the beach into town, where they are to rendezvous with an animal dealer. The dealer has connections with a larger-scale dealer in the capital who is trying to procure some baby chimpanzees to be sold to a shabby circus act in the Middle East. All the men have to do is reach the appointed place and find the man who will pay them the Tanzanian equivalent of fifty dollars for each baby chimpanzee, after which they will blow some of the funds on beer and cheap lodging, and then head back across the great lake to their village homes in the Democratic Republic of the Congo.

The plan goes awry when a policeman on foot patrol notices the shabbily dressed men with the moving sacks as they make their way through a crowded marketplace. The men are detained, the bags opened, and local wildlife conservationists are called. When they arrive at the local police station and open the sacks, the conservation workers find two dehydrated and emaciated baby male chimpanzees. The smugglers explain they were told a baby ape could bring as much as several months' wages if they could make contact with the right people on this side of the lake. There were actually four chimps, they tell the police, but one died during the ten days of traveling, and another was eaten for dinner when the men ran out of food.

So what to do with them? The local government has been cooperating with conservation groups by trying to stop the wildlife smuggling trade. If the baby chimps were not adopted, a bad message would be sent; that breaking up smuggling operations doesn't actually save animal lives, so why not let everyone squeeze a bit of money from the system by allowing the illicit trade to continue? So the orphans are taken in. The two chimpanzees don't look like they'll last long, but they are cared for around the clock by the wildlife workers, and to everyone's surprise they do indeed survive. A month later they are growing, rambunctious infants, eager to spend the day playing, drinking milk formula from bottles, and in general being kids. Once the babies have grown too large to be easily cared for in a household, however, a long-term home will need to be found for them. By the age of twelve they will be difficult to manage, and as they grow into muscular adolescents, their childhood playfulness may morph into aggression toward strangers, making them downright dangerous.

Jane Goodall made a decision long ago to offer to take in confiscated orphans and establish sanctuary homes for them. Many other conservation projects have followed suit, and the result is a network of great ape sanctuaries across Africa and Southeast Asia. The orphaned apes can't ever become wild animals again; chimpanzees taken from the forest and kept even briefly in captivity learn there are easier ways to make a living than foraging far and wide for fruit trees. They inevitably end up approaching human settlements after losing their fear and are shot or trapped. So they spend their early years being raised by a staff of dedicated caregivers and socialize with other such orphans. As they mature they must be banished to large, zoolike cages or established in semi-captive colonies on islands. On such islands ecotourists may visit and learn from them, and the chimpanzees themselves may live in relative freedom.

Today there are chimpanzee sanctuaries across equatorial Africa for displaced orphan chimpanzees, as well as a small number of bonobo sanctuaries in the Democratic Republic of the Congo. There's a fundamental problem with the sanctuary model, one that the system's greatest proponents fully acknowledge, but for which they offer few solutions. Chimpanzees and bonobos live a very long time; in captivity it is not unheard of for a chimpanzee to live age to sixty-five. Maintaining just one chimpanzee in captivity for a year costs thousands of dollars. So running just one sanctuary with thirty or forty rapidly growing apes is an enormously expensive proposition, not to mention a very long-term one.

As I said earlier, we should be far less concerned with the welfare of any one individual animal than with preserving its species in perpetuity. So setting up sanctuaries and running them as ape refugee camps for decades at huge cost may represent misspent funds and human energy. Better to use the money to secure tropical forests in which apes live for future generations and spend one's time and energy preserving ecosystems, which will also protect the apes within. The sanctuary rationale is that if we don't give local African governments a reason to stop smuggling by demonstrating the worth of the lives of individual apes, they won't bother to enforce the laws set up to protect the animals in the first place. It's hard to argue with that. But nonetheless, it's very sad that we may be caring for displaced captives which can never rejoin wild ape societies while the species' natural habitat disappears.

There are various models for ape sanctuaries in Africa. Some are truly care centers, in which large numbers of chimpanzees or bonobos will live for the remainder of their lives. Other sanctuaries attempt to create a semblance of the natural social and environmental needs of the apes. Ngamba Island is a quite successful Ugandan chimpanzee sanctuary in Lake Victoria, a short boat trip from the city of Entebbe. The forty-hectare island was reclaimed from squatters in the 1990s, the human detritus of the place removed, and most of the island left to return to tropical forest, containing many of the plant foods naturally eaten by chimpanzees. Debby Cox, an Australian primatologist, led the early effort along with the Jane Goodall

Institute, and over the years many partner organizations have provided logistical support and funding, with the hope that Ngamba could double as a haven for orphaned apes and pay for itself as a tourist destination. One corner of the island was fenced off as a tourist visitor center, with an elevated walkway and other facilities. As the chimps mature, they are moved into the social group that forages in the forested sector of the island. It's a model for how sanctuaries everywhere might work.

In Asia, orangutans also find themselves orphaned at an appallingly high rate. One study estimated there were more baby orangutans being kept as household pets in Asian countries than in the wild in the 1990s, although confiscations and more widespread public education have largely ended that practice. Orphans are still brought to a handful of sanctuaries in Indonesia, where they are raised and then placed into rehabilitation programs intended to turn them into wild animals again. Tanjing Putting National Park, a swampy, low-lying forest near the southern coast of Borneo in the Indonesian state of Kalimantan, may be the best known of these projects. In addition to being a stronghold for wild orangutans, Tanjing Putting was the site of the original orang rehabilitation center, begun in the 1980s by the orangutan researcher Birute Galdikas. First at Galdikas's main research station at Camp Leakey, and more recently at other sites deep in the park, orangutans have been released in large numbers back into their natural habitat. It's a wonderful feel-good proposition: rescue orphaned orangutans and return them to Nature. But Galdikas's methods raised eyebrows and ecoethical concerns to the point that the Indonesian government took over the rehabilitation operation in the 1990s and continues to run it today. Galdikas had been accused of releasing rehabilitated orangutans into forest already containing populations of wild orangs, a major no-no in conservation circles. Naïve rehabilitants risk aggression from territorial residents, and wild animals face possible disease transmission from captives. In addition to these issues, some observers complained that Galdikas simply had too many orphans to care for them properly or to rehabilitate them successfully.

Wanariset is very different sort of orangutan rehabilitation center, located in East Kalimantan and begun by Willie Smits, a Dutch conservationist. There, in addition to a forestry project that focuses on preserving and managing the natural forest ecosystem, Smits and the Indonesian government have undertaken an orangutan rehab project that emphasizes getting confiscated orphans back into the wild without disrupting natural wild populations.[9] Orphans undergo careful medical screening for pathogens, have limited contact with their human caregivers (unlike Camp Leakey), and are gradually resocialized and released into forest areas from which orangutans have already been extirpated or perhaps have never occurred, but which resemble ideal habitat for any self-respecting orangutan. The goal is to establish, or perhaps re-establish, entirely new wild orangutan populations.

There are now numerous other orangutan sanctuaries in Indonesia, all with the goal of re-establishing the species into forests from which their recent ancestors have been removed. As with chimpanzees, reintroduction works best in areas where the animals no longer exist, so competition for resources and the risk of disease transmission are minimized. And yet there are reasons to be wary of the promise of reintroduction in orangutans. Malaria is known to be more prevalent among rehabilitant orangutans than among wild ones, presumably because the high density at which they tend to be kept increases the infection risk. When released into a forest, they may infect orangutans in neighboring forest tracts, even if care has been taken to release them only into forest not already containing orangutans. Moreover, there has been a notable lack of follow-up regarding the fate of released captive orangutans. We know that orangutans released into existing orangutan territories fare poorly. We don't know if individuals released into forests from which orangutans have been extirpated, or in which they have never lived, fare any better.

Rehabilitated chimpanzees simply cannot be introduced into areas in which there are resident chimps; the residents would attack and attempt to kill such perceived intruders on first contact. Orangutans will tolerate one another's presence, albeit with possible aggression between males. As for bonobos and gorillas, we simply don't know if rehabilitation could work. There are too few animals in sanctuaries in Africa to make such an attempt, and the logistical barriers would be formidable. The bonobos' rarity and their habitat's extreme remoteness make reintroduction an unlikely scenario, and the size of adult gorillas renders them poor candidates for either sanctuary care or reintroduction.

The last sort of sanctuary to consider is the type into which laboratory apes, in the best-case scenario, find themselves after surviving a lifetime of medical experiments or psychological studies. These retirement homes don't excite conservationists, since the chimpanzees in them are nearly certain to remain captives for their entire lives. Whereas apes in sanctuaries in host countries in Africa or Asia have a chance to be repatriated or reintroduced to natural habitat, apes in confinement elsewhere don't, except in extremely rare cases. In the United States, the goal is to give the animals humane lives; a socially and physically enriched environment where they can live out their six-or-so decades. In the case of some of the apes who have been freed from research labs, they enter the world completely unable to socialize normally with others of their species. They are, in fundamental ways, very much like any human prisoner freed from a solitary confinement cell after half a lifetime of captivity. Social skills need to be reacquired, and even basic interaction with the environment—a fresh breeze, the sights and sounds of the outside world—can be very stressful initially. The sanctuaries are refugee camps, but we should be deeply grateful that a cadre of people care enough to dedicate themselves to looking after these unorthodox refugees.

It bothers some in the scientific community to hear apes spoken of as prisoners, or as refugees, as though they were entitled to a life free from human intrusion. This attitude reflects a general reluctance to accept that a few categories of nonhuman animals are so psychologically and emotionally complex that a moral imperative exists to guarantee them the same welfare protection that is promised to children, cognitively impaired adults, and anyone else who cannot speak for him- or herself. This reluctance is considered ignorance when we observe it in the public. But ignorance can be corrected with time and education. The same reluctance among scientists cannot be so charitably attributed. They know better. The main motivation for keeping apes in biomedical research today is the biomedical research industry itself. Chimpanzees are just grist for their mill. Until this mindset is irrevocably changed, we diminish our chances of protecting great apes. More than that, we diminish ourselves.

SIX

The Double-Edged Sword of Ecotourism

Bwindi Impenetrable National Park, Uganda. The gaggle of eco-tourists turns up early on a misty morning at the national park office. They're mostly Americans with assorted other western nationalities—Australian, British, and Italian—in the mix. The Americans look like they stepped off the pages of an L.L. Bean catalog; color-coordinated Gore-Tex ponchos, high-tech boots, and camera bags from which lenses bulge. Their gear costs more than most Ugandans earn in a year. The Aussies are young backpackers, scruffy and tattooed. The British couple are avid birdwatchers, as their distracted up-in-the-trees gaze and $1,500 binoculars attest. The Italians are a chain-smoking man and his girlfriend, who is wearing bright blue spandex and stylish sandals; not exactly the right attire for trekking into rugged, wet mountains. Nevertheless, they're all greeted cheerfully and issued walking sticks by a park ranger, who will serve as the guide. A day earlier, upon entry to the park, each tourist paid five hundred dollars for the privilege of hiking several hours on muddy, tortuous trails, in hopes of spending a carefully timed hour in the proximity of mountain gorillas.

The day begins with a short orientation lecture by the ranger. There are guidelines for conduct to avoid unduly stressing the gorillas and to avoid transmitting our diseases to them. If you turn up for your tour with obvious health issues, like a wracking cough, the guide is supposed to send you home with the offer of a full refund and a rescheduled trip when you're better. Tourists are instructed on how to behave in the face of a charging gorilla—cower and avoid eye contact. They are advised that they will have one hour, not a minute more, in proximity to the gorillas and that they are not guaranteed to get perfect views. On the day in 1995 when I first visited the Bwindi gorillas as a tourist myself, we sat watching the massive, fly-covered rump of a silverback in repose in a thicket. All around were the sounds of gorillas contentedly munching in the undergrowth, but they remained nearly invisible for our appointed hour with them. For the park rangers, it's a trade-off between the opportunity to market the animals as a revenue cash cow while limiting the stress of contact for them as much as possible. At least gorillas live in cohesive groups, so that finding one means you've found perhaps ten or twenty. Chimpanzees don't lend themselves as well to tourism given their fluid, fragmented society; you can easily pick the wrong day or hour and spend it sitting with one somnambulant animal in a tree overhead.

As the group walks out of camp, they pick up an entourage. A knot of soldiers—the Ugandan Patriotic Defense Force, the national army—will accompany the tourists. They look none too excited to be here, their assault rifles dangling casually by their sides. Their presence is the result of the proximity of Bwindi to the international border with the Democratic Republic of the Congo. Only a few kilometers from there, Congolese and perhaps Rwandan rebels may lurk in exile, having melted away into the local villages following the civil wars in those two politically chaotic nations in recent years. There hasn't been a cross-border attack since 1999, and the tourists, most of whom are oblivious to local African politics and the murder of a warden and eight ecotourists here a decade ago, find the sizeable army escort an odd and unwanted distraction.

Once on the trail, all is quiet except for the labored huffing of the chain smoker and the frequent guzzling of bottled water by all. The hike to the gorilla group winds upward out of the camp and, depending on which group is being visited and where it is located, can take from a few minutes to several hours. Sometimes, when the gorillas happen to be very near camp, the guides will lead the group on a circuitous route to give everyone the sense of having discovered them, rather than just walking directly a few minutes from their doors to look at their quarry. The guides stop frequently to allow the novice hikers to rest and to enjoy the spectacular scenery. The British colonists did not call this place the Impenetrable Forest capriciously. The old name reflects the dense green landscape—precipitous peaks covered with high canopy forest, all of which is draped in mats of hanging vines and lush vegetation. It's an easy place to take a fall on the wet, slippery rocks and plant growth, but it's like falling onto a living mattress, with little harm done. There is much to see; it is a bird watcher's nirvana with the largest species diversity of any single forest tract on the African continent, plus monkeys and other forest creatures enough to suit any nature lover.

After ninety minutes of labored walking, the guide puts a finger to his lips and points ahead. The vegetation is flattened in a long trail that heads over the ridge and down the steep slope. The gorillas came this way sometime late yesterday, he says. He has the benefit of knowing where he left the group after the previous day's tourist visit, and since the group only travels a kilometer or so each day, they won't be hard to find now. Not far ahead, in an exposed meadow overgrown with bracken ferns, the guide points out the nest site where the gorillas awoke earlier this morning. There are half a dozen nests scattered over a small area, the largest of which—the silverback's—is enormous, and dotted with piles of his droppings, which are quite enormous themselves. The tourists take turns staring at the bowls of crushed ferns and try to imagine the huge apes going to sleep on this chilly, windswept slope. Then the guide exhorts them to push on.

Only a hundred meters ahead there are branches cracking and bushes swaying, and it's obvious to all that they have reached the gorillas. The guide and his assistants fan out to pinpoint gorillas and try to get tourists into position to see them. With no warning a female charges out of a thicket right at an unsuspecting tourist, who staggers backward until the guide steps up to interpose himself between the gorillas and the terrified onlookers. The female's charge is just a bluff, and she backs into the thicket as quickly as she had shot out of it.

The visitors stand hip deep in greenery, craning their necks for a view of the hidden gorillas, and at first there is little to see. The breaking of vegetation and the contented gurgling of gorilla stomachs is audible to all, as are the grunting contact calls the gorillas make as they forage. But after ten minutes of near-invisibility, a youngster appears, climbing up a liana to inspect the tour group. Eyes widen, camera shutters click wildly. The action lasts only a minute and, after a baleful stare, the baby gorilla drops back into the undergrowth. The tourists are patient, and over the course of the next fifty minutes they're rewarded with the sight of several females and juveniles emerging from thickets and wandering casually past the humans. As the hour closes, the guide checks his watch and signals that their time is up. The tourists turn away and start their hike back to camp, with an over-the-shoulder view of the silverback at last, his domed skull moving regally above the shrubbery. All the way home it pours rain, but the tourists are glowing with the excitement of their brief connection with one of the world's most magnificent creatures.

This is great ape ecotourism at its best. Each group member's five-hundred-dollar payment to spend an hour with wild mountain gorillas, multiplied by the several parties of tourists visiting several different gorilla groups in Bwindi this morning, amounts to more than ten thousand dollars that the Uganda Wildlife Authority has raked in, just in a few hours. In 2009, more than twelve thousand tourists bought Bwindi gorilla permits, making gorilla tourism among the leading revenue sources in the nation of Uganda. For the national park system and the tour operators, it is one of the most lucrative arrangements in the world of ecotourism. The amount paid by each tourist may seem exorbitant, but judging by the wild-eyed look many tourists wear upon returning from their gorilla visit, they have had the experience of a lifetime. Price doesn't seem to be a barrier to high levels of tourism. In fact, the government recently mandated an additional five dollars per day levied on each five-hundred-dollar permit sold by the Uganda Wildlife Authority, which goes to the local economy and helps subsidize new health clinics and dispensaries.

With sixty-four permits available daily (eight tourists per gorilla group times eight habituated groups), a maximum levy of three hundred twenty dollars per day, or about nine thousand dollars per month, can be channeled into the local communities in Kabale and Kisoro districts. This is a point of contention, since the original plan called for twelve percent of the total revenue generated by gorilla tourism to be shared with local villages. A few years later, the passage of the Uganda Wildlife Statute changed the revenue sharing plan to twenty percent of gate receipts (but not other permits). This was in effect a net increase in funding for local people in nearly all the national parks in Uganda, since most of the money generated by safari parks was in the form of park entry fees. Not so for Bwindi and its sister gorilla park Mgahinga, however. There, the levy on gate receipts alone meant local people were cut out of the enormous revenue coming from gorilla trekking permits. However, the revenue is still substantial. About two hundred thousand dollars was raised between 2000 and 2009 to enable the building of at least thirty new schools and ten health clinics and dispensaries in the region.

The funds are intended to compensate local villagers for not cutting down the forest or harassing the gorillas. It works, even in the face of occasional violent incidents, such as a confrontation between a gorilla and a farmer, who finds the huge ape sitting in his banana plantation. That happens because of the single most important element of gorilla tourism: habituation of the gorillas.

HABITUATION

Habituation is a grail for nearly all primatologists. You can only learn so much from wild animals by watching them at great distances or collecting their discarded foods, feces, and DNA. To get inside their societies, you must build trust, and for a smart animal with a history of conflict with the humans living around their lands, trust does not come easily. In the earliest days of research on great apes, the first generation of primatologists used food to entice the animals close enough for observation. Jane Goodall's offerings of bananas brought all the local chimpanzees near but may have also disrupted their natural travel patterns by creating such an intensely desirable food source. Researchers following Goodall have largely abandoned provisioning with food as a habituation strategy. The alternative is to do it all naturally, and that is a long-term, laborious proposition. For a chimpanzee researcher, this can mean many years of frustration spent walking the forest hoping for a glimpse of terrified apes running from you. And because chimpanzees—along with their close cousins the bonobos—live in a fluid, fragmented fission-fusion society, you never have the opportunity to habituate more than a few individuals at a time to your presence, prolonging the process enormously. If the apes in question have a history of being hunted by people, the habituation process can be nearly impossible, entailing years of futility trying to overcome the long-term memory of the animals, who recall the terror that they and their group members experienced.

Gorillas are not like chimpanzees in one important respect. They live in cohesive groups that travel together, eat together, sleep together. And mountain gorillas, because they live in eastern Africa where ape meat is not on the menu for local ethnic groups, have little history of being hunted for food. So the mountain gorilla populations of the Virunga Volcanoes and the Albertine Rift regions of Uganda, Rwanda, and the Democratic Republic of the Congo are ideally situated for tourism. The ease and speed of habituation depend on the personality of the silverback, or silverbacks, in the group. If the group's silverback is laid back, placid and trusting, a wild group may be habituated enough to be followed closely all day long in as little as six months. If, on the other hand, the silverback is a high-strung, flighty male, it means many months of approaching the gorillas only to hear the sound of fleeing apes and find piles of diarrheal feces left behind in fear as they ran. Eventually though, if the animals are approached stealthily but steadily day by day, they will become habituated. Once they lose their flight reaction entirely, tourists can be brought to see them.

DIAN FOSSEY AND ECOTOURISM

There was a time when gorillas were considered so thoroughly savage and dangerous that the idea of paying good money to sit a few yards from them would have been considered insanity. That view became antiquated with the field research of the wildlife biologist George Schaller, whose yearlong study of wild mountain gorillas became the first account of the species and painted a new portrait of a shy, placid, persecuted animal that had far more reason to fear people than vice versa. Then came Dian Fossey. Fossey's pioneering study, in which she adopted the Jane Goodall approach and immersed herself in the society and the daily lives of mountain gorillas, reaped the first truly detailed information about gorillas' intimate lives.[1] She persevered through political turmoil, financial and personal difficulties, and conducted a research project that for eighteen years until her death at the hands of a murderer—likely a local person, perhaps a former assistant, who had watched her end-of-the-shotgun approach to conservation for years.

But Fossey, for all her achievements, was ultimately a failed conservationist. She steadfastly opposed ecotourism for her gorillas. This was ostensibly because she worried about the impact that the tour companies, tourists, and development would have on the animals whose trust she had spent so many years working to gain. Many of her colleagues and students believe that her underlying reason for resisting an invasion of ecotourists was that she enjoyed her privacy in the Virunga Volcanoes and did not want to see her personal fiefdom opened to outsiders. This attitude may have led to her eventual death. The last thing she wanted was tourists walking through the sanctity of her backyard.

Ironically, although she has been lionized by Hollywood in *Gorillas in the Mist*, crucial conservation progress for the Virunga gorillas arrived after Fossey's death. Her former students approached the government of Rwanda with a plan to establish the infrastructure for a small ecotourism program that would bring a great deal of revenue to the cash-strapped local government. This gave Rwandans a greater incentive to protect and cherish their gorillas than Fossey had ever dreamed of. Within a few years, gorilla tourism grew to become the second leading source of foreign cash in the country's economy, trailing only the coffee industry. It remains so today, despite frequent interruptions during Rwanda's chronic bouts of political chaos and civil war. Indeed, during the 1994 genocide, during which perhaps eight hundred thousand people were murdered in a two-month span, only one mountain gorilla died, and that incident was allegedly due to a case of a gorilla in the forest being mistaken for a rebel and shot dead. Both the government and the Hutu militias believed they would end up in charge of the nation, and recognized how vital the gorillas had become for their economic future.

ECOTOURISM AND LOWLAND GORILLAS

Ecotourism works well for mountain gorillas because they are not hunted relentlessly by people, their habitat is relatively accessible, and a visit by jet-setting ecotourists can easily be combined with a safari to the big game parks of East Africa. They are huge and majestic, unlike the much more abundant, smaller, and higher-energy chimpanzee. But ninety-nine percent of the world's wild gorillas are lowland gorillas, and they don't live in eastern Africa. Instead, lowland gorillas are found across a discontinuous band of equatorial African forest tracts, from the eastern edge of the Democratic Republic of the Congo, westward across the vast reaches of the eastern Congo Basin, and finally to the western lowland gorilla strongholds of Gabon, Cameroon, and a tiny corner of Nigeria. Across this area, there are perhaps seven hundred fifty members of the subspecies we call mountain gorillas, and upward of eighty thousand others.

For the most part, gorilla ecotourism means viewing the mountain gorilla populations of East Africa. But numerous ecotourism projects are underway involving western lowland gorillas too. The goal of these projects is to create sustainable development that is compatible with the continued existence of the apes while giving local people a chance to earn a living from the gorillas in their backyard. This would give villagers an incentive not to harm, and even to protect, lowland gorillas.

But the obstacles are many. For one thing, you don't easily link a trip to see lowland gorillas to an itinerary of game viewing. While mountain gorillas live only in fairly accessible tiny forest fragments, they are close enough to the Serengeti, Ngorongoro Crater, and other world-class safari destinations to combine a gorilla visit with a safari in one vacation. Tour operators cater to this crowd; a significant portion of the ecolodges and hotels in mountain gorilla tourism sites are decidedly upmarket, with white-gloved staff serving evening cocktails before guests retire to their luxury tents replete with flush toilets and sinks. Lowland gorillas mostly live in countries lacking in such tourism infrastructure, and indeed lacking in tourists altogether. These areas lack other major wildlife-viewing destinations and the amenities and attractions that draw in high-rolling foreign tourists and their cash. In many areas gorillas are hunted for their meat, so habituation itself is next to impossible. Civil unrest and general lawlessness pervade some of the regions where lowland gorilla tourism could otherwise thrive. Even the landscape is not conducive to ecotourism.

Instead of sweeping panoramas of high mountains and rushing streams, lowland gorillas live in flat, sprawling rain forest that offers no vistas at all. For a relatively few hardy souls, and for local expatriates for whom the journey is not prohibitively difficult, lowland gorillas offer a wonderfully intimate look at creatures of the dark rain forest. Instead of plodding out of a chilly wet meadow like their cousins in the Virungas, lowlanders walk through dimly lit glades and thickets, and approaching to within close observation is a very rare privilege. Big ticket safari-style tourism it is not.

This is not to say that lowland gorilla tourism can never succeed. In the Central African Republic and some portions of the Republic of the Congo, scattered *bais*—swampy saline meadows—attract a multitude of large mammals, including gorillas and forest elephants. Such an open clearing filled with megafauna would be the perfect setting for ecotourism. All that is lacking is the infrastructure to get tourists safely to the site, and of course the tourists themselves. But compared to mountain gorilla ecotourism in East Africa, gorilla tourism in western and central Africa is likely to remain a niche market, and hardly a major source of revenue for any government until infrastructure and political stability improve in the region.

IS ECOTOURISM BAD FOR GORILLAS?

There is a flip side to nearly every benefit of ecotourism. Unforeseen events can cascade until the animal that is the object of loving conservation efforts sees its existence compromised or even threatened by the very same efforts. Ecotourism is a prime example, and great apes may become the victims of their own popularity. From the mid-1990s to the mid-2000s, trips I made to my research camp in Bwindi Impenetrable National Park entailed stopping overnight in the ecotourism center at Buhoma. A sleepy little end-of-the-road village in the early 1990s, Buhoma mushroomed over that decade into a bustling tourist town. There were years when a new tourist lodge appeared on each of my arrivals. At the pace that building was occurring, it would be only a matter of time before some tourism-related health problem would befall the objects of the tourism industry.

In the opinion of some conservationists, the habituation process may harm gorillas and other great apes to the point that disease risks outweigh tourism benefits. It isn't tourism itself that is the problem; it's that great ape tourism usually implies close contact; tourists are allowed to hike right up to wild apes and sit just a few yards away from them on a daily basis. This close approach requires intensive prior habituation by local field assistants, and the daily contact with people continues the process. But such frequent close contact with so many people brings disease transmission risks. Consider a gorilla in Bwindi Impenetrable National Park in one of the groups habituated for tourism. Each day of the year, potentially, up to eight ecotourists plus their entourage of guides, trackers, and armed guards approach them at close range. This means contact with hundreds of people each year. Many of these people are likely to be infected with cold and flu viruses, and at least some each year are no doubt carrying with them more serious transmissible pathogens, from tuberculosis to pneumonia. Because the apes can catch nearly all communicable human diseases, there is serious potential for infections to cross from us to them. And just as with missionaries unwittingly infecting indigenous people, tourists and guides may infect gorillas with no natural immunities to human diseases whatsoever. Wild gorillas and other great apes may thus die from pathogens that are relatively harmless in people who have built up some resistance. Such transmissions have probably occurred many times, and deaths from likely human-transmitted diseases have been recorded. A polio epidemic killed many of Jane Goodall's famed chimpanzees in the 1960s and crippled others. An outbreak of sarcoptic mange, better known as scabies and usually transmitted from human hosts, ravaged the Bwindi gorillas in 2000, thankfully without causing any deaths. Research on the Bwindi gorillas has included investigations into the possible role of humans as vectors of gorilla diseases.[2]

One obvious way to mitigate the risk of contagion from tourists to wild apes involves increasing the safety zone between them. Researchers found that although tourists no longer can approach closely enough to touch the gorillas (although an occasional infant may wander right up to a startled and delighted tourist), the human-gorilla distance is on average about nine feet at closest approach. This is far closer than the seven meters (twenty-three feet) mandated in tourism protocol designed to mitigate disease transmission risk. At some great ape tourism sites, tourists are required to wear surgical masks to limit disease transmission, although the efficacy of this is not yet known. These short distances are hard to avoid when the tourist gorilla groups are so utterly habituated to people that they no longer have any flight distance or fear. Juvenile gorillas are intensely curious about those humans who turn up to gawk at them every day. Limiting the size of tourist groups and the amount of time they spend with gorillas can also be effective strategies. Many of these restrictions are in place already at mountain gorilla tourism sites, but they are routinely disregarded.

Disease transmission due to habituation isn't a given. It depends on the health of the people the apes encounter and the size of the gorilla group (larger groups are more likely to acquire human diseases) as well as the degree to which the apes encounter human diseases in the course of rambling past tourism lodges and camps. The average tourist is likely immunosuppressed by the time he or she has made an intercontinental flight, been dosed with a variety of immunizations to ward off tropical diseases, and has been dealing with all sorts of new intestinal bugs since arrival in Africa. There are rules in place that require tourists to self-report recent medical history and illnesses such as flu and colds before departing on a gorilla tour, with the assurance that a raincheck will be issued. But a tourist on a fixed itinerary hoping for a once-in-a-lifetime opportunity to see wild mountain gorillas is unlikely to admit to being sick. At least one study of chimpanzee-trekking ecotourists in Uganda reported a common prevalence of diarrheal diseases and other infections that few tourists acknowledged before their visit for fear of endangering their spot in the trekking group. Local guides, porters, and guards, meanwhile, are very likely infected with all sorts of local pathogens that haven't been dealt with by the local medical center but to which the people themselves may be highly resistant. When it is introduced to a group of gorillas or chimpanzees, that pathogen can wreak havoc. And these folks have daily contact with the animals, often reaching the gorilla groups before the tourists and departing after the tourists have gone, amplifying the chances of disease transmission.

We know from tourism research that great apes that live in close proximity to people, including those that are routinely visited by tour groups, tend to have higher parasite loads. Gladys Kalema-Zikusoka, a Bwindi veterinary researcher, found with her colleagues that gorilla groups habituated for tourism were infected with the eggs of a variety of parasites that also infect humans, such as *Strongyloides* and *Ascaris*.[3] Similar findings have been made for both chimpanzees and baboons living in and near tourism sites in Tanzania. In 1990, most of the gorillas in tourism-habituated groups in the Virunga Volcanoes became ill with pneumonia, and two animals died. The same sort of respiratory diseases have hit chimpanzees both in tourism sites in Africa and in captivity, and in all cases transmission of human pathogens from local people has been implicated. And having apes living near people obviously exposes them to all manner of fecal pathogens, since sanitation in African village areas is not the best. Moreover, oral pathogens can be passed from food that is spat out, such as once-chewed sugar cane. These can transmit everything from hepatitis to polio plus parasitic worms. All these are from direct transmission from humans to apes via pathogens left on the ground. Malaria and perhaps other infectious diseases that require an intermediate host can also be transmitted.

In addition to all those humans who make scheduled regular contact with the apes, there are numerous chances for unplanned encounters with humans and their detritus. Once habituated, some gorillas quickly realize that it's easier to make a living outside their forest home than in it. If the group's home range is near the forest edge, it's only a short walk into farmers' fields, village areas, refugee camps, ecotourism camps, or the other forms of development that accompany the camps. Encounters with any of these can carry health risks. Along the edges of Bwindi Impenetrable National Park, farmers try to grow bananas, sorghum, maize, and other crops on the precipitous slopes. Gorillas make that a mission impossible. Once habituated, groups travel outside the forest to decimate banana plantations (a chimpanzee would only take the bananas, but gorillas mash and eat the entire tree). They may stay out of the forest, foraging and sleeping for weeks or even months in crop fields. This is not where the average camera-toting ecotourist wants to get his *Gorillas-in-the-Mist* shot. It's also bad for the gorillas. They run a far greater risk of running into a farmer or villager, with a bad outcome for both, or picking up some human-originated bugs in and around villages. Even worse is the contact between the gorillas and the ecotourism camps. Lodges must dispose of their trash, and so trash pits dot the landscape around the tourism center. Once habituated, the gorillas no longer avoid clusters of high human population density. The detritus of people, once ingested and acquired on the feet or hands, can be harmful or fatal to the animals.

Habituated gorillas sit like buddhas, seemingly oblivious to the humans in their midst. But they're not. Just because the animals no longer flee doesn't mean that tourism isn't highly stressful to them. Studies have shown that the gorillas display more stress-related behaviors, and also more aggression, during the hour spent in proximity to a tourist group than they do at any other waking hour of their day. This shouldn't surprise anyone; even for the most mellow silverback and his females, the presence of eight tourists, a guide and his assistants, and an entourage of soldiers creates a presence that cannot easily be ignored. We know all too well that stress can increase the likelihood of disease transmission as well as the development of more serious cases of what would otherwise be manageable illnesses. Stress dents the human immune system just as it does that of every other mammal. Some animals show stress in an obvious way, by seeming agitated or by fleeing. Others don't. A trained eye may be able to see signs of stress in an animal's behavior. Other signs may show up only in stool cultures—gorillas in habituated groups tend to have higher parasite loads than those in unhabituated groups—or in long-term data on disease, reproduction, or survival. It's a very real but not easily detected harmful physical condition.

The habituation process is stressful to great apes. In the beginning, it's all too obvious. They flee, give fear calls, and generally show that the last thing they want is to be habituated. Later in the process, as the apes become accustomed to a human presence, the stress is obvious only when a human moves too quickly or makes a loud noise, sending the tranquil animals into a short-lived panic. But still later, when the apes have come to accept people sitting near them and talking in quiet voices, the stress may still be there. It's just harder to detect. Stress brings about disease outbreaks by lowering immune system defenses, turning subclinical infections into harmful diseases. The sarcoptic mange epidemic in Bwindi may have been brought on by stress; herpes virus is another disease impacted by stress.

The obvious solution, you might think, would be to limit tourism development in order to minimize the harm. This is certainly what most foreign conservation groups argue for. The government, however, is caught between the competing agendas of protecting the gorillas and profiting from them. It is, as in so many other cases, a balancing point between greed and progress. And firmly in the more-is-better camp lie the tour companies, for whom each additional habituated gorilla group is more money in their pockets each year. In this tussle, the bottom line usually wins. At Bwindi there has been a long-term debate over how many groups of the gorillas should be habituated for tourism. The latest 2006 census indicated about three hundred forty Bwindi gorillas, an increase of about ten percent in the past decade. That population is living in about thirty groups, plus a number of lone individuals in the forest. In 2006, five of the thirty groups were habituated. During 2009 more groups were designated to be habituated, so that by 2011 eight groups were habituated to people well enough to be visited by tourists. Approximately twenty-three percent of the gorillas in Bwindi are now habituated for tourism. This is a far smaller percentage habituated than their cousins living in the nearby Virunga Volcanoes, where fully seventy percent of the population is tourism-ready.

It seems clear that the fate of the mountain gorilla is that virtually every individual will someday be habituated to close approach by tourists. This would be the first but probably not the last time that every member of an entire classification unit of a species has been named and monitored. Perhaps this is a sound strategy, scientifically as well as for gorilla tourism commerce. Approachable gorillas are more easily monitored for disease outbreaks, and certainly unhabituated gorillas may suffer epidemics out of sight of the eyes of conservationists. But there's something disturbing about the pressure to treat every last individual mountain gorilla on Earth as an ambassador to ecotourists.

Ecotourism has also been tried numerous times with wild chimpanzees, and it has achieved a moderate level of success, although not with the same enormous monetary potential as with gorillas. For some reason people don't object to plunking down five hundred dollars to spend sixty minutes sitting with gorillas, even though a fraction of that amount gets you close contact with habituated wild chimpanzees. Uganda and Tanzania may be the epicenter for chimpanzee ecotourism. Unlike the case of mountain gorillas, chimpanzee tourism projects are scattered far and wide. Some are connected to famed research sites like Gombe and the Mahale Mountains in Tanzania, where the chimpanzees a tourist can hang out with are the actual animals famed for their roles in the annals of chimpanzee research. In other places the chimpanzees have been habituated expressly for tourism, often near research projects that got the ball rolling years earlier when they opened the area to foreigners with research interests. In other cases, local mom-and-pop habituation projects have gotten the green light from local officials and offer chimpanzee tourism with a hand-scrawled sign sitting on the roadside throughout rural Africa. Chimpanzee ecotourism can work, and although the species isn't nearly as rare as the mountain gorilla, tourism does provide a financial incentive for local people to respect the forests and wildlife, both for the direct benefit of tourist dollars and the indirect benefit of sales of everything from sodas and meals to t-shirts.

We've already seen that the tourism potential for lowland gorillas is far less than the demand for mountain gorillas for a variety of reasons. But the tourism potential for bonobos is dramatically lower than the tourism possibilities for all other great apes. First, they live in areas that are practically inaccessible to all but the most determined scientists. A very long boat trip up-river or a long, expensive small plane charter is only the first step into bonobo territory. Second, bonobos are found in a limited region of rain forest in the heart of Africa, in the Democratic Republic of the Congo. And for most of the past three decades, the DRC has been caught in a never-ending cycle of political turmoil, bloody civil war, and general lawlessness in its remote (and even not-so-remote) regions. This combined with the general rarity of bonobos makes it extremely unlikely that any but the most modest of ecotourism projects is likely to benefit them anytime soon.

Orangutans are also potential targets around which sustainable development via ecotourism can be created. They are the glamour animals in a number of national parks and nature reserves on the islands of Borneo and Sumatra. The risks to the apes from ecotourists are greatly lessened by the habits of the orangutans themselves. Being arboreal, they are meters above the tourists watching them from the forest floor. And being mostly solitary, there is far lower risk of mass contagion from human to ape, or from ape to ape, than there is for group-living gorillas and chimpanzees. Indonesia has an ecotourism industry, but not on the scale of East Africa's by any means, and without the great megafauna of Africa to center wildlife viewing around, Indonesian ecotourists generally go to rain forest sites that are built around observable forest animals, the largest of which are orangutans.

One fundamental difference between ecotourism in Indonesia versus East Africa is that the African safari parks are mostly established in areas where there were no other uses for the land, and not simply because the wildlife was so cherished. "African sleeping sickness," any of several strains of trypanosomiasis, is spread by tsetse flies. It is a widespread disease in sub-Saharan Africa that afflicts both people and their livestock, with domestic hoofed stock and wild game acting as reservoirs of the disease. This kept cattle ranching out of much of the Great Rift Valley, where it otherwise would have competed for space with the great herds of grazing animals. By the time effective medical treatments had been developed in the twentieth century, the exploitation of the region for tourism was already in force.

In Indonesia, there are alternative uses for the forest in which orangutans live that have led to the forests' widespread destruction. As I stated in Chapter 2, conversion of rain forest to palm oil plantations is one of the most dire threats to Indonesian forest conservation, and with it orangutan survival, in that region. Weighing the balance of forest protection versus exploitation by the palm oil industry has not gone in favor of wildlife conservation very often thus far, and the wildlife parks that exist are often on land not suitable for other purposes. There are enough intact forests on Borneo for many reserves holding orangutans to be established and protected, but whether ecotourism can be a driving force behind establishing more reserves is an open question.

PANACEA OR PLAGUE?

Ultimately, ecotourism is neither a cure-all nor a curse for the great apes. In certain well-chosen times and places, it can be a huge boon, as in the case of mountain gorillas, where without tourism the apes' future would be far bleaker than it is. Despite the problems associated with human contact, the gorillas are such a cash cow for the governments and the tourism industries of Rwanda, Uganda, and the Democratic Republic of the Congo that the influx of foreign money almost ensures the apes' future. Not quite, though. One issue that has been solved is loss of habitat. Were it not for tourism, most of the remaining tiny forest patches in which the gorillas live would be already gone. Because of the tourism revenue, the forest cutting has largely stopped, and the gorillas have become a protective umbrella under which myriad other species are given safe haven.

But there is the overarching problem with the business model for great ape ecotourism in the gorilla-holding regions of East Africa. Ecotourism works when western tourists can safely travel to and from a site. But the high-rolling tourists to whom gorilla tourism is oriented are extremely risk-averse, and at the hint of a terrorist attack or civil unrest, they stop coming. So great ape ecotourism is extremely fragile, given that it is entirely based in countries—Uganda, Rwanda, and the Democratic Republic of the Congo—with long histories of political chaos. These days, southwestern Uganda is a safe place, and to ensure that it stays that way, whole encampments of the Ugandan army are stationed near the tourism centers in Bwindi and Mgahinga. Following years of turmoil and repeated civil wars, Rwanda is at relative peace and the gorilla tourists have returned, accompanied by contingents of soldiers. Security in Congo is always a roll of the dice.

Low-budget travelers—backpackers and overlanders—will put up with all manner of hardship, as long as there is a reward at the end of the line—photos to snap and stories to tell back home. But a potential collapse of the ecotourism industry is only one kidnapping or rebel attack away. In early 1999, as gorilla ecotourism at Bwindi was skyrocketing in popularity and its importance to the national economy was reaching its potential, a rebel militia attacked the ecotourism center at Buhoma in Bwindi Impenetrable National Park, Uganda. Actually, calling them a militia is aggrandizement—they were a rag-tag bunch of the same thugs who had carried out the ethnic cleansing in Rwanda several years earlier, and had since been hiding from the Rwandan army in the villages of the eastern Democratic Republic of the Congo. They were remnants of the Interahamwe, militant Hutu extremists who had helped to massacre thousands in the Rwandan genocide. Bwindi's ecotourism center lies within a very short walk from the DRC border, and word had gotten back to the rebels that a raid on Buhoma might net them cash and valuables from western tourists staying in the luxury tented camps there. Although portrayed in the media as a politically inspired assault on Uganda and its western allies, in all likelihood the attack at Bwindi was first and foremost simple banditry that spiraled into something much worse.

A militia of a hundred or more Interahamwe marched from DRC across the border to Uganda, set up a dawn diversionary attack in a nearby town to draw armed guards away from the ecotourism camp at Buhoma, and entered the park gates into the camp. They rounded up all the tourists who had not fled into the forest at the first sound of gunfire, brutally murdered Warden Paul Wagaba to ensure cooperation from the rest, and marched fifteen foreign tourists into the forest. The tourists were divided into several groups, and at some point eight were murdered. Seven others were released, traumatized but physically unharmed. Back in Buhoma, the rebels stole what they could find and burned down some of Buhoma's luxury tented tourist camps.[4] The Ugandan authorities later pursued the rebels into the DRC, arresting and killing some, although it was never altogether clear if those arrested or killed were the perpetrators of the attack.

This sort of civil unrest is fatal to tourism. Visitors to Bwindi, who had numbered in the thousands per year before the attack, dropped to nearly zero for months afterward. For a local economy that has been built on tourism, and which is also the platform for forest conservation in the area, nothing could be worse. The Ugandan government installed a military cantonment in Buhoma, as much to instill a sense of security for tourists as to ward off future attacks. Indeed, after a year, the tourists came back. First came a trickle of backpackers and overlanders, the budget tourists for whom amenities, and apparently personal safety, are held in a delicate balance with cost. Then the luxurious tented camps were rebuilt, and the high-rolling expatriates slowly began to return. Eventually, tourism bloomed again, even bigger than before, and within a few years of the attack, the revenue stream from gorilla tourism was restored, along with the pressures on the gorillas themselves from ecotourists.

The temporary disruption in tourism at Bwindi came about as a result of a single event. An ongoing series of attacks, or an outbreak of full-scale civil unrest in the region, either within Uganda or just across the border, would shut down tourism again in a larger way. Such is the nature of building a conservation program on the wallets of fickle globe-trotting tourists who have multiple options for their once-in-a-lifetime safari experience.

TRUST IN THE APES

There are those who consider the potential costs of tourism to outweigh the benefits, and some of the critics offer alternative proposals to conserve gorillas. All parties acknowledge that if you are going to protect gorillas and the other apes in the long term, you have to offer the host nations and the local people a strong economic incentive to do so. One proposal involves setting up a massive trust fund, donations to which would come from nongovernmental organizations. These organizations could fund-raise in the west for conservation in Africa. Accrued capital could generate interest that could be given to host governments as an inducement, and a reward, for their role in conserving the animals and their habitat. A trust fund would not eliminate the potential for great ape ecotourism. It could, however, mitigate tourism's blowback—the unintended consequences that often result when too many people want close contact with rare animals. With the financial cushion of a trust fund, the local government could take measures to conserve the forest without constantly weighing the competing pressure from tour companies.

Such a trust exists already, on a smaller scale, for both Bwindi Impenetrable and Mgahinga National Parks in Uganda.

An endowment with a charter to aid in the development of local activities, its annual budget is earmarked for community projects that help local people lessen pressure on the forest; on research into the ecology of the gorillas, usually with a potential application to conservation efforts; and for park management activities to help the government do its job of managing the two gorilla populations. The trust (with the acronym MBIFCT) functions independently of the government and of other nongovernmental organizations, but it partners with the various ministries that deal with gorilla tourism, with the local steering committees of villagers, and with the Batwa ("pygmy") tribe that resides near the parks.

MBIFCT performs an important role in gorilla conservation in Uganda. It does not obviate the need for, or the commercial opportunities of, ecotourism. In neighboring Rwanda, a similar trust, named in honor of Dian Fossey, also helps support gorilla conservation activities. Larger organizations also operate all across Africa, a few of them dedicated to great ape conservation. But none can hope to generate enough money to dissuade a government from fostering the growth of great ape tourism. Decreased tourism would lessen the risk of disease transmission from human to ape, and it would minimize the negative impacts of tourism development that we've seen in this chapter. Whether a more comprehensive trust fund is feasible, or just a pipe dream of antitourism conservationists, is an open question.

THE ULTIMATE STAKEHOLDERS

There are a variety of stakeholders in great ape ecotourism. First, of course, are the apes. They can't speak for themselves, so international conservation organizations speak for them and have a place at the negotiating table when tourism policies are decided upon. Second, there are the local people, who ultimately will determine whether the animals survive or not. They can be persuaded with economic incentives to conserve the forest and its inhabitants, as long as political instability is kept at bay. Third is the local government, charged with protecting the apes but also eager to generate income from them. Fourth, the tour companies themselves wield influence in the world of conservation, for better and for worse. Finally, there are the members of the global community who see great apes as cherished natural treasures deserving of protection. And this last faction speaks loudly with its collective wallet.

Great apes can serve an essential role in ecosystem preservation; that of the flagship species. In some tropical forests, protecting top carnivores like tigers or jaguars means creating a vast protected area to serve their enormous ranging needs. These glamour species thus become the umbrella under which countless other species are offered safe haven. Although great apes are not at the top of the food chain, they nevertheless also occupy the role of flagship species, since their substantial habitat needs mean protection for large areas of tropical forest. Their future and the economic future of their host countries are tightly, inextricably linked. Eco-tourism, whatever its costs, is likely to be the path of all nations that are able to take it. The great apes will have to live with the results.

SEVEN

Ethnocide

By comparing the extinction of great apes to a widespread genocide, I mean no disrespect to the millions of humans who have
died inhumanly, or through our collective inaction in the face
of slaughter. Genocide is the willful and systematic destruction
of an ethnic, cultural, political, or racial group. It is a crime in
the eyes of the international community, one that is placed on
an evil pedestal above that of mere mass murder because of its
targeted nature. Groups of people have been trying to wipe out
each other since the beginning of recorded time. Long before
the Holocaust or the Armenian or Rwandan genocides, indigenous peoples were making every effort to wipe one another
from the face of the Earth.

Extinction of the great apes could not be considered a willful act. It would be an act of great willful ignorance. The extinction of a people is not just about the loss of the members of

the society themselves. It's also about the extinction of a culture, which is ethnocide. And great ape culture is in danger of ethnocide. We have spent much time so far talking about the potential loss of ape gene pools. But with every forest that is cut down, we also lose a culture, as surely as the loss of indigenous people means the destruction of their beliefs, language, and religion and all the contributions they had made to the human family.

CULTURE

In the Mahale Mountains of Tanzania, a chimpanzee climbs a palm tree to have a better look around. The forest is dotted with clusters of oil palms, the remnants of trees planted decades earlier when people inhabited villages here. The chimpanzee reaches the crown of the palm, clambers through its frond to sit atop the tree, and surveys his world. Within arm's reach, clusters of the palm fruits sit awaiting his bite, but at no point does he sample them. He's up there simply to enjoy the view. The palms' fruits are coated with an orange layer of thick, fibrous paste, the stuff from which fatty and cholesterol-rich palm oil is extracted. A bounty of fat and carbohydrates goes to waste.

Mahale National Park, Tanzania, comprises a rugged, lush range of hills sloping down to the eastern shore of Lake Tanganyika. It's home to chimpanzees that have been studied by the watchful eyes of generations of research biologists, many of them from universities in Japan. Researchers have recorded tools and other cultural traditions at Mahale that are not seen in the same suite of behaviors anywhere else. Even the chimpanzees of Gombe National Park, a scant hundred kilometers away across the Malagarasi River, exhibit a different set of traditions and tools. Gombe is also prime habitat for oil palms. But a Gombe chimpanzee who climbs an oil palm does so for the express purpose of harvesting its fruits. Palm fruits are in fact the single most widely eaten plant food in the Gombe diet, because the trees produce fruits in every month of the year. They're an invaluable source of fat for an animal whose habitat yields very little of that precious nutrient. Chimpanzees also eat monkeys, wild pigs, and other prey containing animal fat, but unlike those foods, the palms don't flee or bite back. They surrender their treasures readily and risk-free. So why would Mahale chimpanzees fail to capitalize on such a golden opportunity for a free, nutritious meal?

It isn't because they are unavailable or that they're any harder to find or harvest. The chimpanzees at Mahale have simply never learned to make palm fruits a part of their diet. This isn't hard to explain on one level; oil palms are not native to East Africa. They were introduced as commercially valuable trees only in the past century, and they occur only at former village sites. At another level, the spotty occurrence of palm fruit-eating is surprising because they're so nutritious. As with the odd mosaic pattern of tool use, availability is not always the mother of invention for chimpanzees. Clearly, an enterprising and open-minded Gombe chimpanzee stumbled onto palm fruits as a food item decades ago; the dietary choice spread, and now it's an established part of the diet of the entire community. That spark has yet to be provided in Mahale.

If we look farther afield, we find further unpredictability. In Bossou in Guinea, chimpanzees not only eat the fat-rich outer coating of the palm nut; they use their stone tools to crack open the seed itself to get at the highly nutritious kernel. Back in eastern Africa, in Toro-Semliki Game Reserve in Uganda, William McGrew and colleagues noted that chimpanzees do not make use of palms, despite their occurrence.[1] The patchwork pattern of use versus nonuse is apparently random to some extent, in the sense that it doesn't correspond well to variables that primatologists have been able to identify.

In western Africa, chimpanzees in Taï National Park in Cote d'Ivoire came across a piglet in the forest. Their approach scared away the mother pig and the rest of her litter, but this little meal-on-hooves ran in the wrong direction, right into the hands of a hungry male chimpanzee. It's a no-brainer: a tasty snack of raw pork. But the chimpanzee, despite his brief curiosity, did not kill and eat the little piglet. He allowed it to go on its way, unmolested. Back in Mahale, the same encounter between ape and piglet would be quickly followed by squeals of excited delight from the ape, squeals of mortal fear from the pig, and lunch would be served. Why would Mahale chimpanzees pass up a meal that Gombe chimpanzees cherish, and why would Taï chimpanzees disdain a meal that Mahale chimpanzees wolf down? There are no known genetic differences among these chimpanzees that would account for differences in taste. Spicy hot curries are popular in Indian or Thai cuisine but not in traditional Irish or French cooking because of varying cross-cultural traditions, not genetics. The food preferences of wild chimpanzee populations are as subject to the whims of cultural innovation and extinction as are regional human cuisines. What causes one tradition in our societies to catch on and spread is subject to fashion, role modeling, and other vagaries as much as it is to functionality. The same is likely to be true among great apes.

As the great anthropologists of yore went off to remote corners of the world to study nonwestern people, they took with them the desire to see the world from the point of view of the "other." Essential parts of these other worldviews were the rituals in which other people engage and the technologies they employ in their lives. In the days before globalization, the adoption of a particular kind of house construction or a certain form of initiation rite might tell an ethnographer something important about how the "others" were like and unlike us. Bronislaw Malinowski went to the South Seas and challenged our notions that western values about sexuality were universal, laying the groundwork for Margaret Mead's widely publicized later work.[2] Based on his life among Native American tribes, Lewis Henry Morgan argued that cultures could be thought of in evolutionary terms, with units of cultural meaning—what a modern thinker might call memes—that, writ large, make up a society.[3] Alfred Kroeber studied language and kinship terms in order to get at the heart of how indigenous peoples' worldviews are shaped by their languages and vice versa.[4]

Malinowski, Morgan, and Kroeber are regarded as somewhere between quaint and sinister in many modern anthropological circles for having taken such a scientific approach to the study of culture. Nonwestern culture, it is said by cultural anthropologists of the twenty-first century, shouldn't be described by western scientists, since one's personal experience so deeply shapes one's view of others. Only members of the culture itself can and should describe and define themselves. In fact, most modern anthropologists reject the label of social scientist, since it validates an approach in which documenting cultures based on their traditions and tools is a dispassionate, empirical science. This has led to a widespread rejection of the traditional raw material of research, data. Cultural anthropology has tilted away from the social sciences and into the humanities, where often no pretense exists of quantification.

Cultural anthropology no longer seeks to quantify and categorize human cultures as though they were specimens. But primatologists are all about documenting, quantifying, and analyzing patterns of culture in apes. As I said earlier, great apes are the new "savages." We look to them to understand the likely range of behaviors that defined our own deep past. What we find is astounding. They don't radically alter their environments: no spaceships, cell towers, or igloos. They display a very simple version of the same traditions that are the hallmark of extremely intelligent, cognitively complex creatures like ourselves. But the seemingly simple act of passing on behaviors through the generations that are encoded by observational memory, abetted by some trial and error, is preciously rare among living things. It bespeaks a highly developed mind that features forethought and an ability to place oneself inside another's head. Only animals that have big brains and live long lives in the company of others seem capable of making the connections needed to learn such advanced feats of social intelligence.

Cultural traditions among the great apes fall into several categories. First, there are technological traditions: tools. Chimpanzee tool use was first discovered by Jane Goodall at Gombe, and a few years thereafter by Toshisada Nishida and his team in the Mahale Mountains. When Jane Goodall reported tool use in 1961, she was met with widespread skepticism. She saw chimpanzees fashioning probe tools from twigs, inserting them into the tunnels of massive termite mounds, and extracting the tools covered with angry soldier termites that became nutritious snacks. The initial disbelief gave way to a new paradigm in the animal-human continuum as she presented more and more evidence. Nonhuman creatures could clearly possess the foresight and intellect to create something new and useful from the raw materials that Nature provided. Goodall and the other authors of the early tool-use studies assumed that the behaviors were the product of social learning. In the context of field research, where controls don't exist, no one could really show definitively that one animal was copying its tool use from what it observed, while others lacking that opportunity did not. Goodall wisely reasoned that what she saw were cultural traditions, not hard-wired behaviors.[5]

It was a revolutionary perspective. Goodall redefined the animal-human continuum by showing that we were hardly alone in our use of technology. Now, the concept of culture might be extended to nonhuman animals. In the 1970s, she marshaled comparative evidence of culture from the several longer-term chimpanzee field studies available at that time. There were tools and patterns of tool use that were being observed routinely in one chimpanzee community but had never been spotted in another. Other primatologists, notably William McGrew of Cambridge University, continued to compile and analyze cross-cultural data on chimpanzee tool use and other traditions with the goal of understanding the nature of chimpanzee culture. McGrew identified nearly twenty types of tool use across Africa from the seven major field studies at that time.[6]

By the late 1990s, enough long-term field studies of chimpanzees across the African continent had been conducted that a meaningful comparison was possible. Andrew Whiten of the University of St. Andrews encouraged chimpanzee researchers from all the long-term projects to compile their data on cultural traditions, and the results were published in a landmark 1997 paper. Whiten factored out potential environmental issues, such as the presence or absence of termites, or of tool materials that might account for interforest differences in tool use.

He also sought to isolate behaviors, from distinctive grooming techniques to tool construction, that were performed consistently and customarily. Since it's reasonable to assume that genetic differences do not account for why a chimpanzee in West Africa would use a stone tool but not a stick tool whereas the opposite is the case in East Africa, all behaviors left after environmental factors were taken into account would be culturally inherited. That is, they were learned from one's mom, from one's peers, and from the community at large.

Whiten and his colleagues identified at least thirty-nine cultural traditions in chimpanzee societies. They showed that two chimpanzee communities living in forests that are only miles apart may be distinguished on the basis of a difference in consistently seen behaviors. They defined a *tradition* as the use of a given tool or the customary performance of a given behavior. The aggregate of these traditions in one chimpanzee society is a *culture*. This is fundamentally the same way we beancount the myriad patterns of behavior in one human culture to distinguish it from another. Chopsticks versus forks, Mandarin versus English; these plus a hundred other simple dichotomies create a very basic description of Chinese versus western culture.[7]

The cultural traditions identified by Whiten ranged from tool use—tools designed to capture termites and ants—to styles of grooming. They included using leaf sponges to obtain drinking water from tree hollows and leaf grooming, in which a chimpanzee appears to pick at the surface of a leaf randomly plucked from a bush. Although it looks like a purely functional behavior—cleaning off a leaf before eating it—it has a more symbolic purpose. Depending on the forest and the chimpanzee community, leaf grooming can be a symbolic request for grooming from another chimpanzee. It can also get a fellow groomer going again when his enthusiasm for grooming is lagging. Sometimes the ape picks small parasites off the surface of the leaf and squashes them with his finger or against his forearm, which elicits the interest of group mates and may promote a grooming bout. The behavior has been seen among chimpanzees across Africa, from the Republic of Guinea in the far west to Tanzania in the east, but the pattern varies depending on the location. In some places leaf grooming seems more functional, in others more symbolic. Primatologists find the quasi-symbolism of the gesture and its widespread distribution fascinating. Is it possible that the tradition emerged in just one chimpanzee society and then spread across the continent? Or did it pop up simultaneously in many forests through trial and error over many millennia? Is its exact pattern of occurrence due to multiple innovations, or to multiple extinctions of the trait after a formerly widespread distribution? We would love to have answers to these questions, which get at the root of cultural innovation and extinction.

Technological traditions include the termite fishing we've already seen. In addition, chimpanzees fish for ants of several species, primarily obtained from tree trunks. Then there are traditions that don't involve technology: grooming styles, vocal dialects, and some intriguingly symbolic behaviors that leave us wondering what goes on in the mind of the ape. One of Jane Goodall's earliest descriptions of tool use was of leaf sponges. A chimpanzee plucks leaves and chews them into a wadge, then dips the clump into a stream, or into rain water collected in a tree cavity. On a recent visit to Gombe, I watched the venerable female Gremlin sit beside a rushing stream repeatedly dipping a leaf sponge into the current and sucking the cold clear water from the sponge placed in her mouth. All the while her infant watched, trying awkwardly to imitate the behavior. I once watched as Prof, an older male chimpanzee at Gombe, spied a hole in a tree trunk while foraging for fruit. He descended to the ground, searched around until he found a large stick, then carried the stick up the tree and proceeded to use it as a ramrod, shoving it into the hole. At the first plunge, a woodpecker flew out, tipping me off to what Prof had suspected. His next withdrawal of the stick revealed sticky yellow egg yolk dripping from its tip, which he licked with relish. Prof had extended his reach well beyond that of his arm by adopting a tool. Whether he had seen another chimpanzee, perhaps his mother, do this can't be known, but there's little reason to think that observational learning in chimpanzees is any different from that in children.

Using sticks to obtain termites or ants, or honey or eggs, is using a tool to extend the arm's reach. It sometimes involves not one tool but a composite of tools to accomplish a task. A chimpanzee uses one tool to break into a termite mound tunnel, then swaps it for a smaller, finer twig to probe into the tunnel and extract termites. Sometimes the latter tool is itself modified during use, the frayed tip clipped with the teeth to make it more effective. In Bwindi Impenetrable National Park, we recorded chimpanzees using stick tools of varying sizes to forage for the honey (and the larvae, no doubt) of different species of bees. They use a diminutive stick for the ground nests of tiny stingless bees. A longer, stouter stick is needed to break into the tree nests of aggressive honey bees. The longer length of the latter may also have extended the reach of chimpanzees, protecting them from stings.

Sticks and sticklike objects are used in other ways in chimpanzee society. In the Goualougo Triangle in the Republic of the Congo, a long-term study has revealed fascinating new uses of technology. If you've ever used your foot to help drive a shovel or spade into hard soil, you know that a foot-assisted tool can exert far more force than one powered by arms and shoulders alone. Goualougo chimpanzees have also figured that out. A burly male chimpanzee warily approaches a termite mound, unaware that his actions are being caught on a remote camera trap mounted to a nearby tree. He's eager to open the mound's tunnels in search of termites, but the mound's construction from termite castings makes it as hard as concrete. Rather than give up, he picks up a large stick and pushes it against the surface of the mound after first clearing away leaves and debris with his hand. Then, curling his leg up under his chest and placing his foot atop the stick's upper end, he drives his full weight and force into the stick. He has chosen a stick of a particular tree species known for its straight and rigid qualities; he may have carried the long branch-cum-spade from many meters away to this spot. The mound's claylike surface explodes in a powder of reddish dirt; an opening has been made. This termite nest–puncturing behavior has not been seen at any other chimpanzee research site.[8]

Goualougo chimpanzees also use composite tool kits somewhat differently from those described above. Having perforated a termite mound and gained access to a tunnel using a large stick, a chimpanzee will drop that tool and pick up a smaller, more delicate twig, which he or she uses as a probe to fish for the insects inside. They will similarly use a large branch to break open a beehive high in a tree, then follow up with a small stick dipped into the hive to extract honey. If the hole in the tree is not quite wide enough for a hand to reach in, the chimpanzee may employ a third, intermediate-sized tool to widen the cavity. Three tools are used serially, in other words, to accomplish one task. Nothing with this sort of serial-use complexity had been seen before Crickette Sanz and David Morgan observed it in Goualougo.

Musa, an aged female chimpanzee, is ambling along a trail in Gombe National Park when she comes upon a nest of siafu.

Siafu are safari ants of the genus *Dorylus;* they're abundant in many East African landscapes, and they're nasty. From their underground nests they send out vast columns into the forest, eating or killing every small animal in their path. The soldiers, with their half-centimeter jaws, attack fiercely and relentlessly. The bane of every field biologist in East African forests is to be attacked by a swarm of siafu, especially since the modus operandi of the ant is to climb well up inside your pants leg before latching onto some tender skin. Chimpanzees are a lot tougher than I am when it comes to dealing with siafu bites, but they still don't like to be bitten. Yet siafu make a tasty and nutritious snack, and like all insects, offer a major nutritional boost if consumed in quantity.

The problem is getting at them without being bitten savagely. Musa scans the ground for a tool. She picks up a fairly large stick, much longer and more robust than the spindly twigs she might use to fish for termites. She carries this wand into the low-hanging branch of a tree over the ground nest of the siafu. Then, grasping the overhanging branches with both of her prehensile feet and one hand, she uses the other hand to expertly dip into the nest. The ants respond by swarming en masse up the intruding stick, racing toward her hands with mandibles open. Musa waits until the stick is so covered with the brownish ants that it looks like a wand dipped in thick chocolate, then withdraws her tool. She quickly draws the wand through her other hand, ending up with a huge fistful of furious ants, which she plunges into her mouth, crunching them by the hundreds. Although each individual ant is tiny, the sum of calories and nutrients is not insignificant, and is well worth the pain of a few unavoidable bites.

Ant dipping is a cultural tradition that occurs widely across Africa among chimpanzee societies, but with local variations. In some forests ant dipping is a two-handed job. In other forests, especially in West Africa, it's purely a one-handed task performed on the ground next to an ant nest. In a few forests, like Kibale National Park, Uganda—only a few hundred kilometers from Gombe—chimpanzees don't seem to engage in ant dipping at all. The quirky distribution of this and many other tool traditions suggests that cultural innovation, and perhaps also cultural extinction, occur easily and at a rapid rate. Richard Wrangham suggests the rapid rate of innovation may actually be problematic in understanding the spread of chimpanzee culture. We have already seen a number of traditions appear and spread after only a half-century of watching wild chimpanzees. If the frequency of innovation were the limiting factor in the distribution of chimpanzee culture, then innovation would have to be very, very slow, perhaps one new tradition popping up every few thousand years (because we can safely assume that chimpanzees have been around for at least a million years, and have likely been inventing traditions for a very long time). If extinctions of traditions occur just about as often as does their invention, it might account for both the patchy distribution of many traits and also the rapid rate of innovation in spite of many chimpanzee societies lacking this or that tradition.[9]

TOOLS AS WEAPONS

Throwing has been elevated to a high art by humans, from quarterbacks to javelin heavers to baseball pitchers and cricket bowlers. Chimpanzees throw rocks and other objects, but usually underarm and usually not in a very goal-oriented way. In Gombe National Park, researchers are always wary when crossing streams or approaching the waterfall in the company of chimpanzees, because the sound of the water seems to excite the apes. They charge about, and in the course of their excitement they pick up rocks, which they toss underhand with great accuracy. I lost count of the number of times I had to dodge melon-sized rocks heaved in the general direction of my head by Frodo and his followers at Gombe. And there are recent reports of chimpanzees throwing rocks at one another in a seemingly directed manner, as projectiles intended to do harm.

Although used on very rare occasions as projectiles, rocks and stones are far more useful to some chimpanzees in an altogether different way. Chimpanzees in some forests in western Africa collect rocks from the floor of the forest, then carry them to the base of fruit trees. They carefully place a hard-shelled fruit into a well-worn depression atop the exposed roots of the trees, or atop a flat rock, then smash the fruit with the stone.

The ground under such trees may be littered with worn, often-used stone tools. A clearing full of chimpanzees hard at work cracking nuts sounds very much like a busy cobbler's workshop. When Christophe Boesch, a Swiss primatologist, first observed the behavior in Taï National Park in Cote d'Ivoire in the late 1970s, it revolutionized our view of chimpanzee technology. Just when anthropologists had accepted that the definitions of tools and culture had to be broadened to include the modified sticks that chimpanzees use in East Africa, along came an entirely new and unexpected tool culture. Not only do the users gather the stones they need, they appear to heft them to gauge their weight and usefulness for the task at hand. They also lug them for many meters across the forest floor. Most intriguing is the complete lack of a connection between the nature of the environment and the type of tool such chimpanzees use. Gombe National Park in East Africa is a rock-strewn landscape where stones of all shapes and sizes are to be had nearly within arm's reach everywhere. There are plenty of trees in Gombe whose fruits are too hard-shelled to be eaten by the chimpanzees there. Yet Gombe chimpanzees have not puzzled out the value of using stone tools (although they do the reverse, smashing certain tough fruits against the ground or rocks to open them). Conversely, the forests of West Africa where the nut-cracking chimps do their work tend to be lowland rain forests, where appropriately sized stones are few.

The significance of stone tool use by chimpanzees in the forests of western Africa cannot be overstated. Recently, a new generation of archaeologists has extended the search for the beginnings of human material culture to apes. Digging down a meter or two through the forest floor, archaeologists discovered evidence of a stone tool culture that extends back nearly five thousand years. Julio Mercader, an archaeologist from the University of Calgary, works in a field that has customarily seen stone tools made by early humans as the holy grail awaiting discovery. The stones and the nutshells cracked by them are vivid evidence that hundreds of generations of chimpanzees have passed down the culture of tool use from one to the next. How far back chimpanzee ancestors were using tools is impossible to say, but it suggests that once the mental leap is made and tools are invented, they stay in use through the millennia. We know from studies of the human fossil record that the venerable human ancestor *Homo erectus* made and used essentially the same stone tool set for more than a million years. That's some fifty thousand generations. Such technological, and therefore intellectual, conservatism is unthinkable in a modern world in which we line up to buy the newest iPad only months after its predecessor first appears on the shelves. But in a world where technology means survival rather than convenience, you don't casually abandon something that works.[10]

The nut-cracking chimpanzees of Ivory Coast have peers in nearby Guinea. In a research site called Bossou, chimpanzees are expert stone tool users. There, a team of researchers from Japan led by Tetsuro Matsuzawa has studied a community of adept tool users for nearly three decades, with a focus on the finer details of stone tool–using culture. The researchers have not been content with studying naturalistic tool use. They've taken the next step by creating experimental situations in the wild for the chimpanzees to solve. By placing stones strategically near the nut-cracking sites, the researchers have been able to study the ways in which chimpanzees select stones (by size and weight) and how they learn their nut-cracking skills (by watching mom). We know from the work of Matsuzawa and his colleague Tatyana Humle that Bossou chimpanzees tend to display handedness (identifying as right- or left-handed) most strongly when they're engaged in their most cognitively complicated task—nut-cracking. We know that there is a critical developmental window of learning to become an adept tool user from about three to five years of age, and that good tool users "apprentice" themselves to veteran techno-masters, watching them carefully over years in the course of developing the patient adroitness needed to manipulate tools.[11]

An ape-made tool may do more than extend one's reach. By increasing lever arm force, tools have uses that go beyond getting at food and water. To attempt to control female promiscuity, male chimpanzees sometimes use violence, and on rare occasions they turn for help to weapons. For decades we believed that the one purpose for which chimpanzees did not use tools was to fight other chimpanzees or hunt. Despite images from the movie *2001: A Space Odyssey* of early humans braining each other the minute they realized how useful a club can be, wild apes were virtually never seen to do anything like this. But careful analysis of the long-term records of chimpanzee observation from Kibale National Park has revealed a number of instances in which male chimpanzees have brutalized females by wielding branches, usually in the context of male mating ambitions. Female chimpanzees are by nature promiscuous, mating with many males in their community during the course of even one ovulatory cycle. This is no doubt strategic on the part of the females, who cloud paternity in an attempt to make it impossible for any given male to assume he isn't the father of her offspring, thereby decreasing the risk of violence toward her infant. Male chimps try very hard to control female sexuality, to the point of physical intimidation. Martin Muller and his colleagues reported that male chimpanzees mated most often with females to whom they showed the most aggression, and that males were most aggressive to the most reproductively successful females. Sexual aggression among male chimpanzees seems therefore to be linked to the hope of controlling females sexually and reproductively. Are we seeing the evolutionary roots of domestic violence in a forest in Uganda? If we are, the loss of culture that accompanies the loss of every chimpanzee population that falls under the chainsaw or the poacher's gun eliminates another opportunity to get deep inside the origins of human behavior.[12]

As useful as weapons can be when brute force is advantageous, and as obsessed as males have been with weaponry throughout human history, there is precious little evidence that chimpanzees ever employ weapons in hunting. Chimpanzees are mainly fruit eaters, but they are also highly efficient and ruthless predators on other small mammals. During my research on chimpanzee hunting behavior in Gombe National Park, Tanzania, I found that one particularly brave and skilled hunter, the gargantuan male Frodo, could account for the deaths of one-tenth of the study site's red colobus monkeys (the favored prey of chimp hunters everywhere the two species coexist) in a single year. Frodo and the other hunters employed a combination of tree-climbing agility, tactical intelligence, and sheer bravado in capturing monkeys (but never technology). I would marvel at the ability of a hunter to see a situation unfolding in which he (nearly all hunting in Gombe is by males) might capitalize, and then put a plan into operation. For instance, Frodo once waited patiently just below the crown of a palm tree, in the fronds of which sat a group of monkeys. As the monkeys nervously sought an emergency exit from the palm, Frodo gently pulled down a frond, creating the illusion of a passage to safety for the monkeys. As the colobus began to stream across the bridge of fronds, Frodo reached up and lunged at them. This sort of tactical decision-making on the fly is seen by chimpanzees in the act of hunting by anyone who studies their behavior. In some forests, hunting is taken to communal extremes; males appear to cooperate in the hunt, the participation of each enabling the hunting party to score a higher rate of kills than any chimpanzee acting alone could hope to aspire to.[13]

After a half-century observing wild chimpanzees across

equatorial Africa, researchers like to think they understand the basic range of chimpanzee behavior. Chimpanzee social behavior is remarkably uniform wherever we look; most of the local variations are in cultural traditions, tool use, and dietary specifics. So it was with great surprise that we learned recently that chimpanzees in at least one site—Fongoli in Senegal—use sticks to disable or kill potential prey. Among the prey animals eaten by chimpanzees are galagos—small squirrellike nocturnal primates also known as bushbabies for their plaintive nocturnal calls. Galagos spend the day sleeping inside tree trunks. They're out of the reach of a hungry chimpanzee's hand, but not beyond a hand holding a stick. So chimpanzees in Fongoli collect sticks from the ground and carry them up to the tree cavity in which a bushbaby hides. They use the stick as a ramrod to poke the tiny mammal until it is disabled or moves into a position from which it can be captured and eaten.[14]

The media caught wind of this exciting observation and twisted it a bit to suggest that chimpanzees were acting like ice-age hunters, "sharpening" sticks into "spears" and impaling bushbabies like hot dogs at a campfire. It appears that the sticks, while sometimes tapered somewhat by chimpanzees using their teeth, do not impale the prey; they merely bludgeon the animal into immobility. The anecdote I related earlier about watching the Gombe chimpanzee Prof use a stick to poke bird eggs out of a tree hole shows that the behavior pattern seen at Fongoli is hardly unique; only the prey is novel. But the image of a primitive human fashioning a weapon for hunting is hard to keep out of one's head. Perhaps more noteworthy than this isolated occurrence is the rarity of the use of a tool in the context of hunting, when it would seem to be such a useful asset.

NONTECHNOLOGICAL CULTURAL TRADITIONS

Although most of the attention paid to cultural variation among chimpanzee societies from one forest to the next has focused on tools, there are other key ways in which ape cultures are unique. In Gombe in East Africa, a chimpanzee approaches a comrade and sits down beside him. He reaches up and grasps a tree branch just overhead with his right hand, and begins to groom the armpit hair of his buddy with his free left hand. The groomed recipient meanwhile performs a mirror image of the same behavior, one hand on the branch and the other employed in the act of grooming.

Not far away, in Mahale National Park on the same lakeshore, two other chimpanzees are engrossed in the same act. But in Mahale the males are not grasping an overhead branch. Instead, they hold one another's hand. This partnering of grooming is seen only among chimpanzees and bonobos. It may seem like the most mundane evidence of cognitive sophistication. But primatologists consider mutual grooming a cognitively challenging act, not unlike a conversation between two people. It's the small talk of primate communication, a little tactile version of the verbal courtesies that we humans use to reassure one another that the person on the elevator with you, or sitting in the airplane seat next to you, is nonthreatening. Perhaps this level of complexity is why gorillas and orangutans don't do it.

Human cultural traditions prominently and delectably include food preferences. The fiery chiles that would make a Swede gag in disgust are the stuff of which a meal is made in Thailand. Food nourishes us, but far more than that it defines us. Halal and kosher dietary restrictions ban the consumption of pork and shellfish for observant Muslims and Jews. Many tribal societies raise pigs and value pork above all other foods. The roasted spiders advertised for sale at roadside stands in Southeast Asia would not be favorite items in an American suburb. But grilled lobsters—not very different at all from a big aquatic spider—would be relished. The historical roots of dietary taboos and tastes are a mélange of folklore, nutritional need, and food availability. Hindus scorn beef because in ancient times, a live cow was far more valuable than a butchered one. You can only eat the meat once, but the milk, butter, yogurt, and even the urine keep on coming. Once established, dietary traditions, like religion, language, and ethnic costumes, become ways that groups of people separate and elevate their cultural identity from those surrounding them or threatening them. Some Native American tribes still eat puppies at feast celebrations. Anthropologists are convinced that one reason for eating a food so unpalatable to western minds and taste buds is exactly that. It's the one food that will never be co-opted by the dominant society that has closed in on all sides.

Chimpanzees exhibit taste preferences that are often linked to the availability of foods in their habitat. In a forest where no red colobus monkeys live, monkeys in general are rare items on the chimpanzee menu. This is through lack of availability, and because other species may be more difficult to catch, occur at lower density, don't taste as good to a chimp palate, or all of the above. But sometimes the choice of a particular food is not clearly connected to the environment or even runs counter to expectations from ecological factors. For instance, chimpanzees everywhere love meat, and among their meat choices is the flesh of tender young wild pigs. Bushpigs are bristle-haired, tusked, boarlike animals that occur commonly in forests across Africa and are hunted with relish in most places. In Gombe National Park, Tanzania, chimpanzee foraging parties typically stumble across a mother pig with a litter of her piglets hiding in a thicket on the forest floor. Each chimpanzee grabs a squealing piglet and runs for his life while the mother wheels around, snorting and flashing her tusks. A good distance away, the hunters stop to happily eat their lunch of raw pork.

In Taï National Park, Cote d'Ivoire in western Africa, chimpanzees also encounter bushpigs as they patrol the forest. But here, as I related earlier in the chapter, the pigs are not considered a food item. The meat is of the same nutritional value, the chimpanzees stand to benefit every bit as much from a meal of meat as they do in Tanzania. Yet a helpless piglet found on a trail is likely to be left alone. Why would a protein- and fat-starved chimpanzee pass up a meal like this? Eating pigs seems to be nothing more or less than a cultural tradition that popped up and became entrenched in Gombe but not in Taï.

DO ONLY CHIMPANZEES EXHIBIT HIGH CULTURE?

Nearly all of the foregoing discussion has centered on chimpanzee cultures. That's because compared to these apes, the other species' technological skills range from rudimentary to nonexistent. Bonobo researchers are fond of describing the cognitive accomplishments of their ape subjects. In captivity, bonobos and chimpanzees display comparable levels of technological skill. Indeed, some bonobos have displayed remarkable linguistic abilities in laboratory settings. In the wild, bonobos do little in the way of tool use. Male bonobos will drag branches across a forest clearing to signify their intention of getting the group up and moving after an afternoon siesta. But little else of the sort of environmental manipulation that chimpanzees routinely exhibit has been observed. Other cultural traditions are few and far between, in part because so few wild bonobo populations have been studied in detail that no profile can be built to compare how two different bonobo communities might have diverged culturally.

Gorillas have always been thought of as the cattle of the great apes. They're enormously bulky, they're fairly sedentary, and they spend a good deal of their lives munching plants. But recent observations have shown that there is some sophistry to gorilla behavior too. Richard Byrne of the University of St. Andrews has completed years of research on the ways in which subtle differences between gorilla populations in the handling and processing of plant foods may reveal cultural traditions. In the wild, gorillas eat a variety of plants that have thorny or thistly leaves and stinging parts. They eat them despite these protective barriers, but they handle them carefully. In the Virunga Volcanoes, a high percentage of plant foods present some difficulties in handling before eating. Stinging nettles are infamous among gorilla researchers—and the reason you see them wearing gloves in photos—because the plants produce burning rashes at the slightest accidental touch. Young gorillas show obvious discomfort when handling them. Yet the nettles contain nutrition, and if the gorilla can avoid the main stinging parts— the stems, which also contain the most digestion-resistant cellulose—they can be an important part of the gorilla menu. But how to eat a plant covered with stinging parts and not be stung? Gorillas solve this, or at least minimize the pain, by whirling the leaves together with a deft swipe of one hand, turning plant eating into cigar wrapping. With all the leaves of the plant in hand, the offensive leaf-stems are left protruding and can be twisted loose and discarded. The bundle of leaves is then pulled from the closed hand, folded over to put the stinging bristles on the concealed underside of each leaf, and plunged into the mouth.[15]

No one step in this sequence would mean much were it not for the final product: a fairly stingless nettle plant ready to be ingested. We believe the sequence developed culturally, through patient trial and error along with some observation. Byrne identifies four distinct routines in the process of feeding on several noxious plants in separate environments, so it makes little sense to argue that the routine is hard-wired genetically. Not only are individual behaviors learned from one another via observation; whole multi-step sequences of raveling and unraveling stinging leaves are imitated. If we look at other populations of mountain gorillas, we see some variation in the tradition, albeit while working with different species of plants. Just as chimpanzee populations differ in their termiting techniques from site to site, so gorillas appear to vary in their approach to disarming the most unpleasant foods in their habitats.

It may seem that gorilla researchers are grasping at straws in arguing that these slight differences among gorilla societies represent culture in a meaningful way. But to cultural primatologists these differences are as meaningful as any simple differences among human cultures. And they even extend to the Asian orangutan. Although orangutans have never been noted as tool users, researchers have recently made a number of fascinating discoveries about orangutan technology and cultural traditions. In 1996, scientists first reported orangutans in a peat forest in Sumatra using a tool kit; a composite set of sticks for extracting insects from trees. The same orangutans used sticks to assist in prying seeds out of hard-shelled fruits.[16]

These examples of tools would not much impress any chimpanzee or chimpanzee researcher. But when such technologies differ from forest to forest, they suggest elements of simple culture. Carel van Schaik of the University of Zurich and Cheryl Knott of Boston University compared patterns of tool use between their research orangutans in Sumatra and Borneo, respectively. They found that their two orangutan populations showed consistent differences in the ways in which they handled tools used for eating the same fruit species. This suggested learned traditions in technology of the sort that we see in chimpanzees. It was van Schaik's Sumatran orangs that exhibited the most tool use, a fact that van Schaik believes is related to the greater sociality of his apes. Sumatra's peat forests are fertile and highly productive compared to the forests in which orangutans live on nearby Borneo. This productivity appears to allow orangutans in Sumatra to live a more social life in larger party sizes than their solitary Bornean neighbors. More socializing means more opportunities to observe and learn, and this may be critical in the development of tool cultures.[17]

WHAT DOES IT ALL MEAN?

Making sense of the range of cultural traditions in great ape society can be contentious when making the bridge to human culture. I was once attacked at a conference for using the term "cultural diversity" in reference to the remarkable array of traditions we see among chimpanzee societies across Africa. My attacker claimed that "cultural diversity" is a phrase reserved for people, and to apply it to apes would be both inappropriate and demeaning to human ethnic groups. Perhaps the term "behavioral variation" would be appropriate for chimpanzees, she suggested. Whatever we call it, human behavioral variation based on group identity and traditions, and chimpanzee behavioral diversity based on local traditions, are both culture. Within the field of anthropology, the battle over the meaning of culture amounts to a cottage industry as well as a blood feud. Those folks who have staked their career claims on nailing down the definition of culture are not about to share the term with nonhuman animal cultures. At least not without a fight.

To me, a culture is a group of individuals sharing a basic bundle of traditions. The traditions cannot simply be responses to the environment, although certainly some traits of any culture will be. A culture living by the seacoast will likely favor boat builders, and one in the Arctic will probably value skill at killing seals. But many cultural traits are unrelated to or even contradict their ecological context. This is a fundamental aspect of human culture. Some aspects may be functional. Cows are sacred to Hindus because they're worth more alive than dead. But many other cultural traditions and traits are not well connected, if at all, to any function. They were simply invented by someone and picked up by others and spread widely. From there, they may spread within the group, dying out as their inventors and practitioners die, or they may diffuse to other groups as members migrate. Whether the members we're talking about are human or ape doesn't matter. The process is the same.

The destruction of each forest means the loss of one chimpanzee culture. The loss of each chimpanzee culture means we lose a piece of the *meaning* of the species. There is no fundamental difference between the loss of a human culture or language in the face of contact with the outside world and the loss of similar memes among great apes.

My underlying point in this chapter about culture concerns extinction. Cultural traditions are born, and they die. When a forest is razed, the apes and their traditions die with the trees. We fret about the extinction of ancient human cultures that cannot withstand the onslaught of the developed world in Tibet or the Amazon rain forest. Another human language is lost to the world every few weeks. It is the same, albeit on a simpler scale, with the apes. We have no idea how many centuries or millennia it has taken for the cultural traditions we see today to emerge. We know only that, once gone, they are likely not coming back.

EPILOGUE

May There Always Be Apes

It's 2150. The world has changed in ways that no futurist can yet predict. Wars, cataclysms, and epidemics that existed only as science fiction have come and gone. We are connected in ways that make smart phones and the internet look ridiculously primitive. Global population growth, long forecasted to plateau around twelve billion, has yet to level off. The human tsunami has swept clean much of the natural world that in the early twenty-first century we had still hoped to save for posterity. The developing world has been most transformed, as the sprawl of villages and farms has given way to towns and cities, and the forests of the twentieth century have been largely transformed into agricultural land. Indonesia is a developed country now, with cities replacing virtually all of the forest. They produce inexpensive products for use in China, which in the late twentieth century were still being made in China. The forests of Africa have fared somewhat better only because poverty is still so crushing that machine-driven deforestation has not entirely prevailed. In the few-and-far-between patches of forest that remain, many are empty of larger animals. Large mammals and other animals from these former forests must now be kept safely under guard in fenced, patrolled sanctuaries that are little more than large zoos.

And the great apes? For the most part, there are no longer any wild living apes. The orangutans of Sumatra are extinct in the wild, and those of Borneo reduced to a few hundred in each of several forest reserves, where ecotourists fly in every day to pay their respects and ogle them. Mountain gorillas and eastern lowland gorillas are gone too, victims of endless cycles of civil war that ultimately left their tiny forest refuges unprotected at the wrong time, during which soldiers and local people wiped them out. Western lowland gorillas still survive, scattered in the low thousands in pockets of central Africa, but they are so relentlessly persecuted for their meat that no one can hope to get a glimpse of them, let alone establish effective conservation projects. At least their ability to live in degraded forests has allowed them to survive. Chimpanzees have fared worse than lowland gorillas, perhaps due to their need for relatively undisturbed habitat full of fruit trees. Their numbers plummeted during the outbreaks of Ebola that hit central Africa hardest during the mid-twenty-first century and took a further hit from the civil war that split the former Democratic Republic of the Congo into three separate nations. Those calamities, on top of ongoing deforestation and hunting, have made wild chimpanzees one of the most endangered primates in Africa. This is a sad state of affairs for such a versatile ape, one that in 2012 was the most abundant of the remaining species.

In 2150, children still go to the zoo and see great apes, although increasing awareness of the treatment of apes in captivity has resulted in legislation that makes the standard of care so stringent that only the largest and best-funded zoos try to maintain them. Television documentaries still feature apes in both the wild and captivity, as do children's books. And tourists still travel to Africa and sign up for chimpanzee-watching safaris, hoping to get a glimpse of the animals in their own world. But everyone knows how this story will end. If apes are reduced to a total global population of only a few thousand in 2150, by 2250 they will be all but gone from the Earth.

This dystopian image of the future of the apes is terribly bleak. It does not have to happen this way. The contingencies of history have always played a role in the extinction or survival of endangered animals, as fashions of diet and style, and patterns of land use, change. We tend to forget that only a hundred years ago, many of our most common water birds—egrets, herons, and the like—had been threatened with extinction due to a demand for their feathers to adorn broad-brimmed women's hats. Turtle soup nearly spelled doom for diamondback terrapins in the coastal waters of the United States. And the egg-collecting hobby, rampant among earlier generations of amateur naturalists, threatened dozens of species with extinction. The fashion of eating apes today may give way to African cultural traditions in which such a dietary practice is taboo. But don't count on it. Many sources of intervention—such as the formation of the National Audubon Society, which was instrumental in saving birds from plume hunters, or the conservationists who banded together to rescue the last remaining American bison—tend to step in at the last possible moment to save species from permanent extinction. Sadly, that won't work for the great apes. They reproduce so slowly that captive breeding is really not a long-term survival option, other than to maintain our captive apes as captives forever. Only by preserving enough of their habitat can we hope to have wild great apes on Earth a hundred or two hundred years from now. Of course a century or two is a drop in the bucket of a planet's timescale. But it's a far longer time frame than humans are able or willing to plan for, so we must act for the immediate future and hope future generations can be more farsighted than we are.

For some perspective on the issue of extinction, consider the view of our own century that an eighteenth-century visionary might have had. Imagine an American landscape devoid of its vast grassy midwestern plains. Imagine an America in which grizzly bears and wolves live only in a few protected mountainous areas instead of on the plains and valleys of the entire western half of the country. Imagine bison herds no longer roaming Ohio and Pennsylvania, and mountain lions no longer chasing their prey in New Jersey and Connecticut. This is of course the America in which we all now live, and yet we feel we've done a decent job of preserving our wild spaces and our wildlife heritage. The great megafauna of North America, so associated with places like Yellowstone National Park, are in reality animals that naturally lived across most of the current United States. Yellowstone and other Rocky Mountain sanctuaries are just their last refuge following their relentless extirpation from the Great Plains and eastern woodlands. So we must avoid the habit of looking down at Africa and Asia for their ongoing attempts to convert their forests and the animals in them to profit. It's simply what human beings do, and despite a changing ethic about land use, it will continue.

The best-case scenario for the twenty-second century is that there will continue to be viable, well-protected populations of wild great apes in Africa and in Indonesia. There will likely be virtually none living outside those national parks and reserves, and no doubt poaching will still be a problem in some time periods and some places. As great apes decline in numbers and as their host nations develop rapidly, we must hope that we stave off the human onslaught long enough for economic transformations to take place. Tourism infrastructure and political stability bred by increasing development would enable the host countries to view their wildlife as tourism revenue generators rather than food items. With a system of reserves that have real protection, ape populations will survive long term. The mountain gorillas of eastern Africa, reduced to an extremely vulnerable several hundred, are undergoing a small population boom following decades of protection. The same would happen for all great apes given time, habitat space, and freedom from persecution.

As a primatologist, I have often heard conservationists sniping that scientists don't really help to save endangered animals; they just want to study them. It's true that I have plenty of colleagues for whom conservation is not an active concern in the sense that they don't get involved with habitat or wildlife conservation issues directly. But field primate studies have historically been the backbone and foundation of conservation efforts. Long, long before the Gombe or Mahale chimpanzees were recognized as populations in danger of extirpation and in need of careful conservation, they were research subjects. First Jane Goodall and then Toshisada Nishida spent decades carving out successful research projects in remote forests. These are still research sites today, but they also function as carefully protected gene pools and as ecotourism centers. Dian Fossey fiercely protected her own privacy along with the sanctity of her mountain gorillas in Rwanda. After her death, her followers engaged the government in a new map of how research and conservation could work. Today, the mountain gorillas of the Virunga Volcanoes are perhaps the most iconic and most carefully monitored primate population on Earth. Once again, field research gave way to a conservation effort. Conservation and research are synergistic, not at odds. In the modern world, one should never take place without the other.

Ultimately, whether or not great apes survive the coming centuries will depend on the active involvement of the countries in which they live. All the well-intentioned conservation energy and money in the world will not save orangutans if the government of Indonesia, either openly or covertly, allows its forests to be lumbered for the sake of plantations and the disposable chopstick industry. Just as nineteenth-century American ranchers and land owners resented European economic meddling in the new western states and territories, the ape-holding nations of the developing world should be expected to allow their wildlife, including their great apes, to slide into extinction unless the western conservationists provide an economically compelling path to preserving them. In the democratic, economically functioning ape-holding countries, there is reason for optimism. In those countries in which the rule of law and an infrastructure do not exist—such as the largest nation in which apes live, the Democratic Republic of the Congo—we can only wait and hope for conditions to improve soon enough that grassroots conservation efforts can function in collaboration with a healthy government.

The outlook for great apes is bleak, but only fatal if one assumes a worst-case scenario for the coming hundred years. Despite deforestation (both legal and illegal), poaching, and disease, the great apes have hung on. The coming decades will decide their fate. In a world in which poverty, bankrupt governments, and political instability are the rule, they are doomed. In a world in which the continued survival of the apes is seen as a benefit to all involved, they will survive. Their future is very much ours.

Notes
Further Reading
Acknowledgments
Index

Notes

ONE *Heart of Darkness*

1. F. Boas, *The Mind of Primitive Man* (New York: MacMillan, 1911).

2. J. Conrad, *Heart of Darkness*, first published as a three-part novella in *Blackwood's Magazine*, 1899.

3. The Great Ape Project New Zealand, Inc., *Submission on the Animal Welfare Bill (No. 2)*, Oct. 1998; "MPs launch charter for animals," *The Star* (Christchurch, New Zealand) 25, Oct. 9, 1980.

4. J. Goodall, *The Chimpanzees of Gombe: Patterns of Behavior* (Cambridge, MA: Harvard University Press, 1986).

5. C. Boesch and H. Boesch, "Optimization of nut-cracking with natural hammers by wild chimpanzees," *Behaviour* 3 (1982): 265–286.

6. C. Sanz and D. Morgan, "Chimpanzee tool technology in the Goualougo Triangle, Republic of Congo," *Journal of Human Evolution* 52 (2007): 420–433.

7. Goodall, *Chimpanzees of Gombe*.

8. Researchers at Ngogo, Uganda, have witnessed both inter-community killing and intracommunity killing. The latter is what one

might call panicide, the ape equivalent of homicide. D. P. Watts, "Intracommunity coalitionary killing of an adult male chimpanzee at Ngogo, Kibale National Park, Uganda," *International Journal of Primatology* 25 (2004): 507–521.

9. When I began my work on chimpanzee hunting in Gombe National Park in the early 1990s, my colleague Janette Wallis and I undertook an analysis of the older data collected to see what patterns would emerge. These papers led to a 1998 monograph on the predator-prey ecology of chimpanzees and red colobus at Gombe. C. B. Stanford, J. Wallis, H. Matama, and J. Goodall, "Patterns of predation by chimpanzees on red colobus monkeys in Gombe National Park, Tanzania, 1982–1991," *American Journal of Physical Anthropology* 94 (1994): 213–228. C. B. Stanford, *Chimpanzee and Red Colobus: The Ecology of Predator and Prey* (Cambridge, MA: Harvard University Press, 1998).

10. T. Nishida, "A quarter century of research in the Mahale Mountains: An overview," in *The Chimpanzees of the Mahale Mountains*, ed. T. Nishida, 3–36 (Tokyo: University of Tokyo Press, 1990).

11. C. Becquet, N. Patterson, A. C. Stone, M. Przeworski, and D. Reich, "Genetic structure of chimpanzee populations," *PloS Genetics*, April 2007.

12. D. Fossey, *Gorillas in the Mist* (Boston: Houghton Mifflin, 1983).

13. S. L. Clifford, N. M Anthony, M. Bawe-Johnson, K. A. Abernathy, C. E. G. Tutin, L. J. T. White, M. Bermejo, M. L. Goldsmith, K. McFarland, K. J. Jeffrey, M. W. Bruford, and E. J. Wickings, "Mitochondrial DNA phylogeography of western lowland gorillas (*Gorilla gorilla gorilla*)," *Molecular Ecology* 13 (2004): 1551–1565.

14. The original papers on captive bonobo social and sexual behavior grabbed headlines when published. The view from the ground in their natural habitat provided a somewhat different picture. F. B. M. de Waal, "Tension regulation and nonreproductive functions of sex in captive bonobos (*Pan paniscus*)," *National Geographic Research Reports* 3 (1987): 318–335. See also C. B. Stanford, "The social behavior of

chimpanzees and bonobos: empirical evidence and shifting assumptions," *Current Anthropology* 39 (1998): 399–420.

15. G. Hohmann and B. Fruth, "New records on prey capture and meat eating by bonobos at LuiKotal, Salonga National Park," *Folia Primatologica* 79 (2007): 103–110; Stanford, "The social behavior of chimpanzees and bonobos."

16. The degree to which bonobos and chimpanzees truly differ has been questioned by an increasing number of researchers, as summarized in this article by a group studying bonobos in Planckendael, a zoo and research facility in Belgium. J. M. G. Stevens, H. Vervaecke, and L. Van Elsacker, "The bonobo's adaptive potential: social relations under captive conditions," in *The Bonobos: Behavior, Ecology, and Conservation*, ed. T. Furuichi and J. Thompson, Developments in Primatology: Progress and Prospects (Zurich: Springer, 2008), 19–38.

17. C. P. van Schaik and P. van Duijnhoven, *Among Orangutans: Red Apes and the Rise of Human Culture* (Cambridge, MA: Harvard University Press, 2004).

18. *Morotopithecus* is currently considered the oldest ape in the fossil record, although it certainly did not look much like a modern ape. D. L. Gebo, L. MacLatchy, R. Kityo, A. Deino, J. Kingston, and D. Pilbeam, "A hominoid genus from the early Miocene of Uganda," *Science* 276 (1997): 401–404.

TWO   *Homeless*

1. United Nations Environment Programme, "The great apes —the road ahead: a GLOBIO-perspective of the impacts of infrastructural development on the great apes." UNEP-World Conservation Monitoring Centre, Cambridge, 2002, available at www.globio. info.

2. See, for example, this page of the Home Depot website: https://corporate.homedepot.com/CorporateResponsibility/Environment/WoodPurchasing/Pages/FAQs.aspx. Accessed May 2012.

3. See, for example, this website about retail use of the tropical

hardwood nyatoh tree (*Payena sp.*, Family Sapotaceae): http://www
.rainforestrelief.org/What_to_Avoid_and_Alternatives/Rainforest
_Wood/What_to_Avoid_What_to_Choose/By_Tree_Species/Trop
ical_Woods/N/Nyatoh.html.

4. See, for example, K. Foerstel, "China in Africa," *CQ Global Researcher* 2 (Jan. 1, 2008): 1–26.

5. http://www.worldwildlife.org/who/media/press/2009/WWF Presitem12414.html; http://www.rnw.nl/english/article/green-facade-asia-pulp-paper; see also R. A. Butler, L. P. Koh, and J. Ghazoul, "REDD in the red: palm oil could undermine carbon payment schemes," *Conservation Letters* 2 (2009): 67–73.

6. http://www.worldwildlife.org/who/media/press/2009/WWF Presitem12414.html.

7. Peter Gelling, "Forest loss in Sumatra becomes a global issue," *New York Times*, Dec. 6, 2007.

8. S. A. Wich, I. Singleton, S. S. Utami-Atmoko, M. L. Geurts, H. D. Rijksen, and C. P. van Schaik, "The status of the Sumatran orang-utan *Pongo abelii*: an update," *Oryx* 37 (2003): 49–54.

9. Peter Aldhous, "Borneo is burning," *Nature* 432 (2004): 144–146; see also I. Singleton et al., *Orangutan Population and Habitat Viability Assessment: Final Report*, IUCN/SSC Conservation Breeding Specialist Group, Apple Valley, MN, 2004.

10. D. Morgan and C. Sanz, *Best Practice Guidelines for Reducing the Impact of Commercial Logging on Wild Apes in West Equatorial Africa* (Gland, Switzerland: IUCN/SSC Primate Specialist Group, 2007).

11. Forest Stewardship Council: http://www.fsc.org.

12. C. A. Chapman, S. R. Balcomb, T. R. Gillespie, J. P. Skorupa, and T. T. Struhsaker, "Long-term effects of logging on African primate communities: a 28-year comparison from Kibale National Park, Uganda," *Conservation Biology* 14 (2000): 207–217.

13. Morgan and Sanz, *Best Practice Guidelines.*

14. C. P. van Schaik and P. van Duijnhoven, *Among Orangutans: Red Apes and the Rise of Human Culture* (Cambridge, MA: Harvard University Press, 2004).

15. W. F. Laurance, T. E. Lovejoy, H. L. Vasconcelos, E. M. Bruna, R. K. Didham, P. C. Stouffer, C. Gascon, R. O. Bierregaard, S. G. Laurance, and E. Sampaio, "Ecosystem decay of Amazonian forest fragments: a 22-year investigation," *Conservation Biology* 16 (2002): 605–618.

16. Ibid.

THREE *Bushmeat*

1. E. Bowen-Jones, *The African bushmeat trade: a recipe for extinction. Report for the Ape Alliance* (Cambridge: Fauna and Flora International, 1998).

2. Mark Kinver, "Illegal bushmeat 'rife in Europe,'" BBC News, June 17, 2010. Available at http://www.bbc.co.uk/news/10341174.

3. A. Plumptre, personal communication.

4. R. A. Mbete, H. Banga-Mboko, P. Racey, A. Mfoukou-Ntskala, I. Nganga, C. Vermeulen, J. L. Doucet, J. L. Hornick, and P. Leroy, "Household bushmeat consumption in Brazzaville, the Republic of the Congo," *Tropical Conservation Science* 4 (2011): 187–202.

5. Dale Peterson wrote what is still the definitive account of the bushmeat trade in Africa and its conservation impact. D. Peterson, *Eating Apes* (Berkeley: University of California Press, 2003).

6. D. S. Wilkie and J. F. Carpenter, "Bushmeat hunting in the Congo Basin: an assessment of impacts and options for mitigation," *Biodiversity and Conservation* 8 (1999): 927–955.

7. Peterson, *Eating Apes.*

8. Wilkie and Carpenter, "Bushmeat hunting."

9. K. Ammann and J. Pearce, *Slaughter of the Apes: How the Tropical Timber Industry Is Devouring Africa's Great Apes* (London: World Society for the Protection of Animals, 1995).

10. J. R. Poulsen, C. J. Clark, G. Mavah, and P. W. Elkan, "Bushmeat supply and consumption in a tropical logging concession in northern Congo," *Conservation Biology* 23 (2009): 1597–1608.

11. Ammann and Pearce, *Slaughter of the Apes.*

12. Wilkie and Carpenter, "Bushmeat hunting."

FOUR   *Outbreak*

1. M. Bermejo, J. D. Rodríguez-Teijeiro, G. Illera, A. Barroso, C. Vilà, and P. D. Walsh, "Ebola outbreak killed 5000 gorillas," *Science* 314 (2006): 1564.

2. Ibid.

3. C. Mann, *1491: New Revelations of the Americas before Columbus* (New York: Vintage, 2006).

4. J. Goodall, *The Chimpanzees of Gombe: Patterns of Behavior* (Cambridge, MA: Harvard University Press, 1986).

5. A. E. Pusey, M. L. Wilson, and D. A. Collins, "Human impacts, disease risk, and population dynamics in the chimpanzees of Gombe National Park, Tanzania," *American Journal of Primatology* 70 (2008): 738–744. J. M. Williams, E. V. Lonsdorf, M. L. Wilson, J. Schumacher-Stanley, J. Goodall, and A. E. Pusey, "Causes of death in the Kasakela chimpanzees of Gombe National Park, Tanzania," *American Journal of Primatology* 70 (2008): 766–777.

6. S. Kondgen, H. Kuhl, P. K. N'Goran, P. D. Walsh, S. Schenk, N. Ernst, R. Biek, P. Formenty, K. Matz-Rensing, B. Schweiger, S. Junglen, H. Ellerbrok, A. Nitsche, T. Briese, W. I. Lipkin, G. Pauli, C. Boesch, and F. H. Leendertz, "Pandemic human viruses cause decline of endangered great apes," *Current Biology* 18 (2008): 260–264.

7. Pusey et al., "Human impacts;" Williams et al., "Causes of death."

8. P. D. Walsh, T. Breuer, C. Sanz, D. Morgan, and D. Doran-Sheehy, "Potential for Ebola transmission between gorilla and chimpanzee social groups," *American Naturalist* 169 (2007): 684–689.

9. F. Gao, E. Bailes, D. L. Robertson, Y. Chen, C. M. Rodenburg, S. F. Michael, L. B. Cummins, L. O. Arthur, M. Peeters, G. M. Shaw, P. M. Sharp, and B. H. Hahn, "Origin of HIV-1 in the chimpanzee *Pan troglodytes troglodytes,*" *Nature* 397 (1999): 436–441.

10. A. Chitnis, D. Rawls, and J. Moore, "Origin of HIV Type 1 in colonial French Equatorial Africa?" *AIDS Research and Human Retroviruses* 16 (2000): 5–8.

11. B. F. Keele et al., "Increased mortality and AIDS-like immunopathology in wild chimpanzees infected with SIVcpz," *Nature* 460 (2009): 515–519.

12. C. A. Chapman, S. R. Balcomb, T. R. Gillespie, J. P. Skorupa, and T. T. Struhsaker, "Long-term effects of logging on African primate communities: a 28-year comparison from Kibale National Park, Uganda," *Conservation Biology* 14 (2000): 207–217.

13. C. E. Hawkins, C. Baars, H. Hesterman, G. J. Hocking, M. E. Jones, B. Lazenby, D. Mann, N. Mooney, D. Pemberton, S. Pyecroft, M. Restani, and J. Wiersma, "Emerging disease and population decline of an island endemic, the Tasmanian devil *Sarcophilus harrisii*," *Biological Conservation* 131 (2006): 307–324.

FIVE   *In a Not-So-Gilded Cage*

1. The Great Ape Project New Zealand, Inc., *Submission on the Animal Welfare Bill (No. 2)* (Oct. 1998); "MPs launch charter for animals," *The Star* (Christchurch, New Zealand) 25, Oct. 9, 1980.

2. J. Yee, "US legislators seek ban on chimp experiments," *BioEdge*, Apr. 28, 2011, available at http://www.bioedge.org/index.php/bioethics/bioethics_article/9503/. For opposing views see R. P. Bartlett, "Stop using chimps as guinea pigs," *New York Times*, August 10, 2011; J. Cohen, "Scientists decry proposed ban on chimp research," *ScienceInsider*, Aug. 5, 2010, available at http://news.sciencemag.org/scienceinsider/2010/08/scientists-decry-proposed-ban-on.html.

3. M. Haederle, "Animal rights groups face off with scientists over fate of chimps," *Los Angeles Times*, Sept. 3, 2010.

4. Ibid.

5. Animal Welfare Act, U.S. Department of Agriculture, National Agricultural Library, available at http://awic.nal.usda.gov/nal_display/index.php?info_center=3&tax_level=3&tax_subject=182&topic_id=1118&level3_id=6735&level4_id=0&level5_id=0&placement_default=0.

6. S. Inoue and T. Matsuzawa, "Working memory of numerals in chimpanzees," *Current Biology* 17 (2007): R1004–R1005.

7. R. Fouts and S. T. Mills, *Next of Kin* (New York: William Morrow, 1997).

8. D. J. Povinelli, *Folk Physics for Apes: The Chimpanzee's Theory of How the World Works* (Oxford: Oxford University Press, 2003).

9. "Borneo Orangutan Survival (Australia)," Wanariset Orangutan Reintroduction Project, available at http://www.orangutans.com.au/Orangutans-Survival-Information/Wanariset-.aspx.

SIX   *The Double-Edged Sword of Ecotourism*

1. D. Fossey, *Gorillas in the Mist* (Boston: Houghton Mifflin, 1983).

2. M. H. Woodford, T. M. Butynski, and W. B. Karesh, "Habituating the great apes: the disease risks," *Oryx* 36 (2002): 153-160.

3. G. Kalema-Zikusoka, J. M Rothman, and M. T. Fox, "Intestinal parasites and bacteria of mountain gorillas (*Gorilla beringei beringei*) in Bwindi Impenetrable National Park, Uganda," *Primates* 46 (2005): 59-63.

4. C. B. Stanford, "Gorilla warfare," *The Sciences* (July/August 1999): 18-23.

SEVEN   *Ethnocide*

1. W. C. McGrew, L. F. Marchant, C. Payne, T. Webster, and K. D. Hunt, "Chimpanzees at Semliki ignore oil palms," *Pan Africa News* 17,2 (2010): 19-21.

2. B. Malinowski, *Sex and Repression in Savage Society* (New York: Routledge, 1927).

3. L. H. Morgan, *Ancient Society* (New York: H. Holt, 1877; repr. Tucson: University of Arizona Press, 1985).

4. A. Kroeber, *The Nature of Culture* (Chicago: University of Chicago Press, 1952).

5. J. Goodall, *The Chimpanzees of Gombe: Patterns of Behavior* (Cambridge, MA: Harvard University Press, 1986).

6. William McGrew's body of work on chimpanzee tool use and

tool cultures is required reading for anyone interested in the subject. For example, W. C. McGrew, *The Cultured Chimpanzee* (New York: Cambridge University Press, 2004).

7. Whiten et al. (1999) is the seminal paper taking a systematic look at chimpanzee cultural variation. It marked the first time the principal investigators of the several longest-term chimpanzee field studies got together to pool their data in an effort to elucidate large patterns. A. Whiten, J. Goodall, W. C. McGrew, T. Nishida, V. Reynolds, Y. Sugiyama, C. E. G. Tutin, R. W. Wrangham, and C. Boesch, "Cultures in chimpanzees," *Nature* 399 (1999): 682–685.

8. C. Sanz and D. Morgan, "Chimpanzee tool technology in the Goualougo Triangle, Republic of Congo," *Journal of Human Evolution* 52 (2007): 420–433.

9. R. W. Wrangham, "Chimpanzees: the culture-zone concept becomes untidy," *Current Biology* 16 (2006): R634–R635.

10. T. Humle and T. Matsuzawa, "Laterality in hand use across four tool-use behaviors among the wild chimpanzees of Bossou, Guinea, West Africa," *American Journal of Primatology* 71 (2009): 40–48.

11. Ibid.

12. M. N. Muller, S. M. Kahlenberg, M. Emery Thompson, and R. W. Wrangham, "Male coercion and the costs of promiscuous mating for female chimpanzees," *Proceedings of the Royal Society, Series B* 274 (2007): 1009–1014.

13. C. B. Stanford, *Chimpanzee and Red Colobus: The Ecology of Predator and Prey* (Cambridge, MA: Harvard University Press, 1998).

14. This study generated a good deal of controversy. The media wanted to characterize the chimpanzee hunters as cavemen sharpening their spears for a big game hunt. It seems much more likely that the chimpanzees simply tapered their sticks and then jabbed them into bushbabies to stun them rather than spear them. J. D. Pruetz and P. Bertolani, "Savanna chimpanzees, *Pan troglodytes verus*, hunt with tools," *Current Biology* 17 (2007): 412–417.

15. R. W Byrne, "Clever hands: the food processing skills of mountain gorillas," in *Mountain Gorillas: Three Decades of Research at*

Karisoke, ed. M. Robbins, P. Sicotte, and K. Stewart, 294–313 (Cambridge: Cambridge University Press, 2001).

16. E. A. Fox and I. Bin Muhmammad, "New tool use by wild Sumatran orangutans (*Pongo pygymaeus abelii*)," *American Journal of Physical Anthropology* 119 (2002): 186–188.

17. C. P. van Schaik and C. D. Knott, "Geographic variation in tool use on *Neesia* fruits in orangutans," *American Journal of Physical Anthropology* 114 (2001): 331–342.

Further Reading

ONE *Heart of Darkness*

Mittermeier, R. A., N. Myers, J. B. Thomsen, G. A. B. da Fonseca, and S. Olivieri. "Biodiversity hotspots and major tropical wilderness areas: approaches to setting conservation priorities." *Conservation Biology* 12 (1998): 516–520.

Pimm, S. L., G. J. Russell, J. L. Gittleman, and T. M. Brooks. "The future of biodiversity." *Science* 269 (1995): 347–350.

Raven, P. H. "The politics of preserving biodiversity." *Bioscience* 40 (1990): 769–774.

Reid, W. V. "Tropical deforestation and species extinction." In *How Many Species Will There Be?* ed. T. C. Whitmore and J. A. Sayer, 55–73. London: Chapman and Hall, 1992.

TWO *Homeless*

Fimbel, R., A. Grajal, and J. G. Robinson. "Logging and wildlife in the tropics: impacts and options for conservation." In *The Cutting Edge: Conserving Wildlife in Logged Tropical Forests*, ed. R. Fimbel, A. Grajal, and J. G. Robinson, 667–695. New York: Columbia University Press, 2001.

Harrison, M. E., S. E. Page, and S. H. Limin. "The global impact of Indonesian forest fires." *Biologist* 56 (2009): 156–163.

Hockings, K. J., J. R. Anderson, and T. Matsuzawa. "Use of wild and cultivated foods by chimpanzees at Bossou, Republic of Guinea: feeding dynamics in a human-influenced environment." *American Journal of Primatology* 71 (2009): 636–646.

Johns, A. D. "Species conservation in managed tropical forests." In *Realistic Strategies for Tropical Forest Conservation*, ed. J. Sayer and T. C. Whitmore, 15–53. Gland, Switzerland: IUCN—The World Conservation Union, 1992.

Johns, A. D., and J. P. Skorupa. "Response of rain forest primates to habitat disturbance: a review." *International Journal of Primatology* 8 (1987): 157–191.

Meijaard, E., and D. Sheil. "A logged forest in Borneo is better than none at all." *Nature* 446 (Apr. 26, 2007): 974.

———. "The persistence and conservation of Borneo's mammals in lowland rain forests managed for timber: observations, overviews and opportunities." *Ecological Research* 23 (2008): 21–34.

Meijaard, E., and S. Wich. "Putting orang-utan population trends into perspective." *Current Biology* 17 (2007): R540.

Plumptre, A. J., and A. G. Johns. "Changes in primate communities following logging disturbance." In *The Cutting Edge: Conserving Wildlife in Logged Tropical Forests*, ed. R. Fimbel, A. Grajal, and J. G. Robinson, 71–92. New York: Columbia University Press, 2001.

Plumptre, A. J., and V. Reynolds. "The effects of selective logging on the primate populations in the Budongo Forest Reserve, Uganda." *Journal of Applied Ecology* 31 (1994): 631–641.

Rao, M., and C. P. van Schaik. "The behavioral ecology of Sumatran orangutans in logged and unlogged forest." *Tropical Biodiversity* 4,2 (1997): 173–185.

Skorupa, J. P. "Response of rain forest primates to selective logging in the Kibale Forest, Uganda." In *Primates: The Road to Self-Sustaining Populations*, ed. K. Benirschke, 57–70. New York: Springer, 1986.

Wilkie, D. S, F. E. Shaw, G. F. Rotberg, G. Morelli, and P. Auzel. "Roads, development, and conservation in the Congo basin." *Conservation Biology* 14 (2000): 1614–1622.

Wilkie, D. S., J. G. Sidle, and G. C. Boundzanga. "Mechanized logging, market hunting, and a bank loan in Congo." *Conservation Biology* 6 (1992): 570–580.

THREE   *Bushmeat*

Alvard, M. "Testing the 'ecologically noble savage' hypothesis: interspecific prey choice by Piro hunters of Amazonian Peru." *Human Ecology* 21 (1993): 355–387.

Alvard, M. S. "Conservation by native peoples: prey choice in a depleted habitat." *Human Nature* 5 (1994): 127–154.

Formenty, P., C. Boesch, M. Wyers, C. Steiner, F. Donati, F. Dind, F. Walker, and B. Le Guenno. "Ebola virus outbreak among wild chimpanzees living in a rain forest of Côte d'Ivoire." *Journal of Infectious Disease* 179, Supp. 1 (1999): S120–S126.

Koster, S. H., and J. A. Hart. "Methods of estimating ungulate populations in tropical forests." *African Journal of Ecology* 26 (1988): 117–127.

White, L. J. T. "Biomass of rain forest mammals in the Lope Reserve, Gabon." *Journal of Animal Ecology* 63 (1994): 499–512.

Wilkie, D. S., B. Curran, R. Tshombe, and G. A. Morelli. "Managing bushmeat hunting in the Okapi Wildlife Reserve, Democratic Republic of Congo." *Oryx* 32 (1998): 131–144.

FOUR   *Outbreak*

Caillaud, D., F. Levréro, R. Cristescu, S. Gatti, M. Dewas, M. Douadi, A. Gautier-Hion, M. Raymond, and N. Ménard. "Gorilla susceptibility to Ebola virus: the cost of sociality." *Current Biology* 16 (2006): R489–R491.

Leendertz, F. H., F. Lankester, P. Guislain, C. Neel, O. Drori, J. Dupain, S. Speede, P. Reed, N. Wolfe, S. Loul, E. Mpoundi-Ngole, M.

Peeters, C. Boesch, G. Paul, H. Ellerbrok, and E. M. Leroy. "Anthrax in western and central African great apes." *American Journal of Primatology* 68 (2006): 928–933.

Lonsdorf, E. V., D. Travis, A. E. Pusey, and J. Goodall. "Using retrospective health data from the Gombe chimpanzees study to inform future monitoring efforts." *American Journal of Primatology* 68 (2006): 897–908.

Pinzon, J. E., J. M. Wilson, C. J. Tucker, R. Arthur, P. B. Jahrling, and P. Formenty. "Trigger events: enviroclimatic coupling of Ebola hemorrhagic fever outbreaks." *American Journal of Tropical Medicine and Hygiene* 71 (2004): 664–674.

FIVE   *In a Not-So-Gilded Cage*

Bailey, J. "An assessment of the role of chimpanzees in AIDS vaccine research." *Alternatives to Laboratory Animals* 36 (2008): 381–428.

———. "An examination of chimpanzee use in human cancer research." *Alternatives to Laboratory Animals* 37 (2009): 399–416.

Hare, B., J. Call, B. Agnetta, and M. Tomasello. "Chimpanzees know what conspecifics do and do not see." *Animal Behaviour* 59 (2000): 771–785.

Inoue, S., and T. Matsuzawa. "Working memory of numerals in chimpanzees." *Current Biology* 17 (2007): R1004–R1005.

Knight, A. "The poor contribution of chimpanzee experiments to biomedical progress." *Journal of Applied Animal Welfare Science* 10 (2007): 281–308.

Smits, W. T. M., Heriyanto, and W. Ramono. "A new method of rehabilitation of orangutans in Indonesia: a first overview." In *Proceedings of the International Conference on "Orangutans: The Neglected Ape,"* ed. J. J. Ogden, L. A. Perkins, and L. Sheeran, 29–40. San Diego: Zoological Society of San Diego, 1994.

Wolfe, N. D., W. B. Karesh, A. M. Kilbourn, J. Cox-Singh, E. J. Bosi, H. A. Rahman, A. T. Prosser, B. Singh, M. Andau, A. Spielman. "The impact of ecological conditions on the prevalence of malaria among orangutans." *Vector Borne Zoonotic Diseases* 2 (2002): 97–103.

SIX   *The Double-Edged Sword of Ecotourism*

Adams, H. R., J. Sleeman, and J. C. New. "A medical survey of tourists visiting Kibale National Park, Uganda, to determine the potential risk for disease transmission to chimpanzees (*Pan troglodytes*) from ecotourism." In *Proceedings of the American Association of Zoo Veterinarians 1999*, ed. C. K. Baer, 270–271. American Association of Zoo Veterinarians, Media (Philadelphia), 1999.

Adams, H. R., J. Sleeman, I. Rwego, and J. C. New. "Self reported medical history survey of humans as a measure of health risk to chimpanzees (*Pan troglodytes schweinfurthii*) of Kibale National Park, Uganda." *Oryx* 35 (2001): 308–312.

Butynski, T. M., and J. Kalina. "Gorilla tourism: a critical look." In *Conservation of Biological Resources*, ed. E. J. Milner-Gulland and R. Mace, 294–313. Oxford: Blackwell Science, 1998.

Eberle, R. "Evidence for an alpha-herpes virus indigenous to mountain gorillas." *Journal of Medical Primatology* 21 (1992): 246–251.

Kalema, G. "Epidemiology of the intestinal parasite burden of mountain gorillas in Bwindi Impenetrable National Park, S.W. Uganda." *Zebra Foundation, British Veterinary Zoological Society Newsletter* (Autumn 1995): 18–34.

Macfie, L. "Case report on scabies infection in Bwindi gorillas." *Gorilla Journal* 13 (1996): 19–20.

McNeilage, A., M. M. Robbins, M. Gray, W. Olupot, D. Babaasa, R. Bitariho, A. Kasangaki, H. Rainer, S. Asuma, G. Mugiri, and J. Baker. "Census of the mountain gorilla population in Bwindi Impenetrable National Park, Uganda." *Oryx* 40 (2006): 419–427.

Sandbrook, C., and S. Semple. "The rules and reality of mountain gorilla tracking." *Gorilla Journal* 34 (2007): 14–17.

Wallis, J., and D. R. Lee. "Primate conservation: the prevention of disease transmission." *International Journal of Primatology* 20 (1999): 803–826.

Weber, W. "Primate conservation and ecotourism in Africa." In *Perspectives on Biodiversity: Case Studies of Genetic Resource Conservation and Development*, ed. C. S. Potter, J. I. Cohen, and D. Janczewski,

129–150. Washington, DC: American Association for the Advancement of Science Press, 1993.

SEVEN *Ethnocide*

Breuer, T., M. Ndoundou-Hockemba, and V. Fishlock. "First observation of tool use in wild gorillas." *PLoS Biology* 3,11 (2005): e380.

Gruber, T., Z. Clay, and K. Zuberbühler. "A comparison of bonobo and chimpanzee tool use: evidence for a female bias in the Pan lineage." *Animal Behaviour* 80 (2010): 1023–1033.

Sanz, C., C. Schöning, and D. Morgan. "Chimpanzees prey on army ants with specialized tool set." *American Journal of Primatology* 71 (2009): 1–8.

Epilogue

Pusey, A. E., L. Pintea, M. L. Wilson, S. Kamenya, and J. Goodall. "The contribution of long-term research at Gombe National Park to chimpanzee conservation." *Conservation Biology* 21 (2007): 623–634.

Wrangham, R. W. "Why the link between long-term research and conservation is a case worth making." In *Science and Conservation in African Forests*, ed. R. W. Wrangham and E. Ross, 1–8. Cambridge: Cambridge University Press, 2008.

Acknowledgments

This book was much longer in the making than in the writing. When field primatologists sit around a table in the evening in some forest in Africa or Asia and drink beer and tell tales, they are often of the intractable problems facing their study subjects. We bemoan the latest poacher's theft, illegal tree-felling, and host country bureaucratic or societal woes that prevent effective conservation measures from being enacted. A visitor sitting at the table with us might get the impression that we are cynical about the odds of actually saving the apes. Indeed, I know a few colleagues who are quite pessimistic about the coming century. But most of us have cautious optimism, and it is to those people that I have dedicated this book. I even get to thank a few of them personally here.

The idea of writing a conservation-oriented great ape book took shape in the early years of my postdoctoral field project in

Gombe National Park, Tanzania. Being small and surrounded by a human sea, Gombe's conservation issues are vividly on display to even the casual observer. Later, during my decade in and around Bwindi Impenetrable National Park, Uganda, I witnessed an entirely different set of issues; a grand wilderness park being eaten away by development, tourism, and political instability. But in both places, ardent efforts have been underway for many years. My primary debt at Gombe is of course to Dr. Jane Goodall, whose pioneering role and continuing vision have enriched the worldview of more people than all other primatologists combined. She has carried Gombe Stream Research Centre into its sixth decade as the longest-running study of wild animals ever conducted. Goodall and her corps of dedicated conservationists and researchers have transformed the prospects of the forest, the animals, and even the local people of the Gombe area through community-based conservation efforts. Anthony Collins has helped to guide Gombe research and conservation through several decades as well. The heroes of Gombe may be the Tanzanian research assistants, who understand the daily lives of wild chimpanzees better than nearly anyone alive. The foreign researchers come and go, but without the Tanzanian field assistants, research at Gombe could not continue. During my years at Gombe, the team was headed by the late Hilali Matama, and included Eslom Mpongo, Hamisi Mkono, Yahaya Almasi, Selemani Yahaya, Gabo Paulo, Bruno Herman, the late Msafiri Katoto, Issa Salala, the late David Mussa, Karoli Alberto, Tofficki Mikidaddi, Madua Juma, Nasibu Sadiki, and Methodi Vyampi. I am also grateful to the offices of Tanzania National Parks (TANAPA) and the Tanzanian Commission for Science and Technology (COSTECH) for their kind permissions for those years.

In 2010, I made a brief trip back to Gombe, my first visit there since the mid-1990s. It was a chance to reunite with the chimps and the people and also to show my three children why the place held my fascination for so long. It provided an opportunity to see how Gombe might have changed over the past decade. I'm grateful to the current generation of park staff and especially Tourism Warden Lameck Matungwa and Dr. Shadrack Kamenya for helping with logistics.

In Uganda, my work could not have been completed without the dedicated support and help of a large number of people, Ugandans and expatriates. First and foremost, my former doctoral student Dr. John Bosco Nkurunungi was an invaluable collaborator, first as a research assistant, later as a student advisee and co-worker, and today as an esteemed colleague and friend. Equally important was our head field assistant, Gervase Tumwebaze. I have had the pleasure of watching Gervase grow from a young man I hired as a field assistant in 1996 to a locally important politician and pillar of the Bwindi-Nkuringo community today. He is a shining example of how local people living near a forest reserve can become wonderful ambassadors for conservation efforts.

For his initial invitation in 1995 to work in Bwindi Impenetrable National Park, I thank Dr. Eric Edroma, former director of (as it was called at the time) Uganda National Parks. For permission to work in Uganda from 1996 through 2005, I am grateful to the Uganda Wildlife Authority (UWA), the Ugandan National Council for Science and Technology (UNCST), and the Institute of Tropical Forest Conservation (ITFC). In and around Bwindi, I was helped by Drs. Alastair McNeilage, Richard Malenky, Martha Robbins, Nancy Thompson-Handler, Michele Goldsmith, Wardens Christopher Oreyema and Keith Masana, Mr. Simon Jennings, Robert Berygera, Johanna Maughn, Fenni Gongo and his father, Mzee Gongo, Senior Ranger Silva Tumwebaze, Ambrose Ahimbisibwe, and many others. I thank the men of the ITFC staff for their assistance and friendship in the forest and in a variety of permanent and temporary camps over the years. At Camp Kashasha, my project was enabled by the dedicated hard work of Gervase Tumwebaze, Caleb Mgambaneza, and Evarist Mbonigaba, plus a team of local men and women recruited as needed by Gervase. Mitchell Keiver played a key role in the project, and for that and the hardship he endured during the rebel attack in 1999, he has my lifelong gratitude. In the aftermath of that incident, the office of the United States Information Service and Mr. Virgil Bodeen were of tremendous help in getting information back to the United States about the situation as it unfolded. For their help in managing and analyzing the data back at the University of Southern California, I thank Tatyana White, Robert O'Malley, Adriana Hernandez, and Xuecong Liu.

For their help in either reading and critiquing parts of the manuscript, or for pointing me to information that I had not discovered on my own while writing the book, I thank Drs. Roberto Delgado, Erik Meijaard, Ian Redmond, Ian Singleton, and Serge Wich. Three anonymous reviewers also provided much-needed corrections and additions.

This is my fourth book with Harvard University Press and, as always, I am indebted to Michael Fisher for his support and insights. I also thank Elizabeth Knoll for her helpful advice and support, and the editing staff at Harvard University Press. And as always, I thank Russell Galen of Scovil, Galen, and Ghosh Literary Agency for encouraging me to continue doing what I love to do.

My three children have watched me write books for fifteen years now, and have reached the ages at which they can be among my best critics. So I thank Gaelen, Marika, and Adam. And as always, I thank my wife, Erin, for a lifetime of support and love.

Index